Pompeii,
Latrines and Down Pipes

Pompeii, Latrines and Down Pipes

A General Discussion and Photographic Record of Toilet Facilities in Pompeii

Barry Hobson

Assisted by Helen Molesworth and Kate Trusler

BAR International Series 2041
2009

Published in 2016 by
BAR Publishing, Oxford

BAR International Series 2041

Pompeii, Latrines and Down Pipes

ISBN 978 1 4073 0468 7

BAR Publishing is the trading name of British Archaeological Reports (Oxford) Ltd. British Archaeological Reports was first incorporated in 1974 to publish the BAR Series, International and British. In 1992 Hadrian Books Ltd became part of the BAR group. This volume was originally published by John and Erica Hedges Ltd. in conjunction with British Archaeological Reports (Oxford) Ltd / Hadrian Books Ltd, the Series principal publisher, in 2009. This present volume is published by BAR Publishing, 2016.

Printed in England

BAR
PUBLISHING

BAR titles are available from:

BAR Publishing
122 Banbury Rd, Oxford, OX2 7BP, UK
EMAIL info@barpublishing.com
PHONE +44 (0)1865 310431
FAX +44 (0)1865 316916
www.barpublishing.com

ACKNOWLEDGEMENTS

The author wishes to acknowledge a great debt to the many people who have helped in the production of this book. Firstly to the combined efforts of a large number of people associated with the Anglo-American Project in Pompeii - especially the Directors, Drs. Rick Jones and Damian Robinson and the environmental specialists Drs. Jill Thompson and Andrew Jones. Thanks, also, are due to the many students over the years whose enthusiasm has been a source of amazement and encouragement and whose help in the 'cleaning' of latrines has been much appreciated. Special thanks must go to Helen Molesworth and Kate Trusler both of whose phenomenal ability to spot down pipes is unsurpassed, and also David Rider, Emily Schurr and Sue Trusler. Without the assistance given to me by these helpers I would never have been able to survey what has amounted to almost the whole of the city of Pompeii. I am also indebted to the Soprintendenza of Pompeii and his staff for permission to carry out the research and for the cooperation of the *custodi* who gave me access to parts of the city which are not usually open to public view.

Other people must be thanked for their generous support, especially all those who attended the symposium on Roman Latrines in Rome in 2008 including Gemma Jansen (whose article on latrines in 'Sequence and Space' was the first I read on the subject), Ann Koloski-Ostrow, Miko Flohr and Adam Goldwater.

Finally I am deeply grateful to my wife who has allowed me to spend six weeks each summer in Pompeii for the last thirteen years.

CONTENTS

INTRODUCTION

In 1996 the author was privileged to start working, as a student, with the Anglo-American Project in Pompeii, which was carrying out extensive excavation in Regio VI, Insula 1. This association continued until excavation was completed in 2006. During this time a number of latrines, lying in the insula, were recorded, this research resulting in a Master's Degree (Hobson, B. 2004). For comparison with VI.1 the whole of Regio VI was explored, the latrines photographed and certain inferences drawn. It soon became obvious that, in order to be certain about these suppositions, the whole of the city should be surveyed. This coverage of the city has taken a number of years and although, unfortunately, there are a number of omissions, mainly of buildings where conservation and repair are taking place, a comprehensive inventory now exists. Included within this is a record of the down pipes, which indicate the presence of upper storey latrines and the Appendix to this volume contains a photographic record of the latrines and down pipes in the whole of the excavated regions of the city of Pompeii.

Apart from exploration and cleaning i.e. the removal of modern covering material, two main sources of information have been used to explore the existence and whereabouts of the latrines. These are 'Gebaudeverzeichnis und Stadtplan der antiken Stadt POMPEI' by Liselotte Eschebach and 'La Descrizione di POMPEI per Guiseppe Fiorelli' by Umberto Pappalardo. These are acknowledged throughout, as also are other publications which make reference to latrines. In a number of the cases which these quote, the degree of evidence of the existence of the latrines is minimal. This is because the superficial observation of the properties has a number of drawbacks which could only be avoided by carrying out excavation. Despite these problems over two hundred latrines have been identified and a similar number of down pipes from upper floors, producing considerable evidence to indicate that the ordinary inhabitants of the city were exceptionally well provided for, with respect to toilet facilities.

The identification of the latrines has been based upon a number of observable factors. Firstly infrastructure:

1. Pedestals
2. Slots in the walls
3. Plaster marks on the walls
4. Floor tiles and drain holes
5. Windows
6. Niches
7. Supporting arches

Secondly, the position in the property

1. Kitchen
2. Near to door and outer wall
3. Under staircase
4. Upper storey latrines and down pipes
5. Garden
6. In bath areas
7. Underground i.e. in cellars
8. On a balcony

Most of the latrine construction technology is of mixed stone and mortar. However, brick construction, probably in the first century CE, is identifiable in one case (VI.5.4). The lack of preservation of wood has meant that seat material and design is hypothetical. But it is likely that they would have resembled later stone or marble latrines. Latrines with two pedestals are the easiest to recognise[1], but those with only one pedestal and a slot in the opposite wall as in VI.7.15 make it possible to recognize those with two slots[2]. Once these have been identified there is a logical move to the plaster marks on the walls[3]. These have a special relevance if they finish at a height of around fifty centimeters which appears to be the standard height of the wooden seating.

Floor tiles in a narrow room constitute very circumstantial evidence especially if a drainage hole can be seen. Nearly twenty-five percent of the latrines have obvious tiles. This relative low number is because part of the modern conservation practice of the site involves spreading floors with a layer of fine gravel. This renders a degree of invisibility which can only be corrected by scraping off this layer to identify the tiles. The results of this cleaning is quite gratifying (See VII.1.45).

In Pompeian buildings windows are rare with most of the rooms facing inwards towards open areas such as *atria* or *peristyles.* Small rooms with windows which open onto the street are particularly associated with, and almost diagnostic of latrines.[4] Room size, lighting and decoration are important factors for consideration. An assumption is made that most, if not all of the rooms housing latrines would have been dark and that light would be provided from a lamp, sometimes housed in a niche in the wall (Mygind 1921:319). There is very little evidence for this. Incidentally there was a lamp, found in the excavation of the cesspit on the ramp to VI.1.1 (Hobson 2002), perhaps dropped accidentally into the latrine. A small number of the latrines surveyed in Regio VI were in

---

[1] **I.7.1: I.9.11.12a: I.9.11,12b: I.11.10: I.11.11: I.11.14: I.11.15. I.11.17: I.12.2: I.12.11: I.14.2: I.14.5: I.17.2: II.4.10: II.8.1: II.8.3: II.8.5: V.4.1: V.4.3: VI.1.22: VI.1.18,20: VI.5.8,20: VI.6.8,22: VI.7.23: VI.9.6,9: VII.2.18: VII.4.48: VII.10.9,12: VIII.3.12: VIII.4.7: IX.1.20,30:**

[2] **Others are: I.15.1: II.1.10: II.2.1,2,3: V.2.4: VI.3.26,27,28: VI.7.15: VI.7.20,21,22: VI.7.25: VI.8.12,13: VI.10.14: VI.11.8: VI.13.12,19: VI.16.17: VII.1.32:VII.2.16: VII.4.48: VII.9.47,48,51,65: IX.6.1,a: IX.8.3,6,a: IX.9.1: IX.9.11**

[3] **E.g. VI.2.10:**

[4] **I.6.15: I.12.7: II.4.10: II.8.3: VI.2.9,10: VI.2.16,21A: VI.4.4: VI.9.6,9: VII.2.20,21,41B: VII.11.4,5: VII.15.4: VIII.7.20,21,27,30**

rooms which are very gloomy (VI.7.20,21,22: VI.9.6: VI.12: VI.15.8,9). In addition there are some latrines in extremely dark underground cellars.

Niches are particularly helpful when recognising upper storey latrines[5], and there are one or two niches at ground floor level. The drainage from the latrines into cess pits imposes considerable stress onto the nearby walls. To obviate this, supporting arches can be seen[6].

Kitchen areas are frequently the sites for latrines[7]. This may be an historical development from other Mediterranean cultures. However a larger number of latrines are positioned near doorways with drainage into cess pits under the sidewalks of the streets[8]. This is noteworthy in some of the *insulae* in Regio VI. Usage of space and perhaps privacy may explain why some latrines are situated under staircases[9]. This may apply where stairs are immediately accessible from the street.

A considerable portion of the city is open to gardens and orchards. Many of the properties, particularly in the south west of the city have *triclinia* outside so that diners could take advantage of the pleasant climatic conditions. Latrines have been identified in these gardens[10]. Very occasionally the latrine was placed in the peristyles area (e.g. VI.16.7,38) and of course in association with baths. It might be expected congregations of

---

[5] **I.7.1,a: I.11.8: I.12.7: V.1.8: V.1.30: V.1.32: V.3.10: VI.2.27: VI.7.19: VI.8.7,8: VI.8.12,13: VIII.3.16: IX.1.31; IX.2.8: IX.2.19,21,A: IX.6.e:**

[6] **E.g. VIII.4.46: VIII.3.12**

[7] **I.6.15: I.7.7: I.12.12: II.2.1,2,3: II.4.20: VI.1.9,10,23: VI.1.18,20: VI.2.9,10: VI.2.27: VI.5.9,19: VI.6.1,8,22: VI.7.23: VI.725: VII.1230-32: VIII.3.16: IX.3.14**

[8] **1.6.15: I.11.1,2A: I.11.8: I.12.11: I.14.2: I.14.8,9: II.1.7,8,9B: II.4.10: II.8.1: V.1.23,25,26,10: V.4.1,2: V.1.5: VI.1.9,10,23: VI.1.18,20: VI.2.16,21B: VI.2.26: VI.58,20: VI.5.9,19: VI.6.1,8,22: VI.7.4,5,6: VI.7.15: VI.7.19: VI.7.20,21,22: VI.9.3,12: VI.9.6,12: VI.11.8: VI.13.10,11: VI.13.12,19: VI15.4,5,24,25: VI.16.17: VIII.3.12: VIII.3.14: VIII.3.17: IX.2.19,21B: IX.5.11,13: IX.6.1,a: IX.7.21,22: IX.8.1: IX.8.3,6,a: IX.9.1:**

[9] **I.4.26: V.2.4: VI.14.40: VII.4.36: VII.5.23:**

[10] **E.g. I.5.2: I.11.1,2: II.1.7,8,9: II.8.1:II.8.2,3: VI.1.1:**

people would place a demand for latrines. There is a multiple seat latrine near to the amphitheatre in the Grand Palaestra, and a smaller one by entrance to the large Theatre.

Latrines in the upper stories of the city have long been recognized[11], but no research has been carried out on the many down pipes which are visible in walls throughout the city. Visible from the Via Stabiae the latrine in V.1.30 can be seen as a niche in the wall above the beam holes for the floor of the first storey. The horizontal part of the pedestal for the wooden seat is well defined and the terracotta pipe leading down from the latrine can be seen in the corner of the walls. The property at VII.12.14, called the School House, has a narrow room with two down pipes. One of these is coming from under the seat of the first storey latrine, whilst the other appears, from the scar in the wall, to be coming from a second storey latrine. This interpretation may explain a number of dual down pipes running together in walls in other parts of the city (e.g. IX.1.9 and IX.3.15.). Running in the walls of the city there are many terracotta pipes, the upper ends of the vast majority of which are broken and give no hint of where they are from. Down pipes can be seen in different parts of the properties and their existence is said to confirm upper floors (Pirson 1999:170). In order to understand the use of the upper storeys of the buildings in Pompeii it is extremely important to understand exactly what function these pipes had. Two main functions can be postulated for these pipes. Firstly, they may transmit water from roofs, for drinking or other purposes, or they may be carrying effluent material from upper storey latrines. The illustration below shows the flange which secures a tight fit from the upper pipe into the one beneath it. This type of construction will only allow the passage of material downwards and will prevent seepage from the junction. If clean water is to be transmitted this might not be important, although leakage might be a nuisance. However other building features may also have played a part in reducing any problems such as leakage and smell which might

[11] **V.1.8: V.1.30: V.3.10: VII.2.18: VII.2.20; VII.12.14: VIII.4.37: VII.6.7:IX.2.7,8: IX.2.19,21: IX.3.4: IX.3.17: IX.5.21; IX.13.1**

be associated with a pipe which is transmitting faeces and urine. One such obvious building technique is enclosure of the pipe within plaster (e.g. VII.13.25: VIII.7.5)) or stone construction (e.g. VII.3.35:VII.4.38: VII.9.50: VII.12.13: VIII.5.10: IX.3.14). Occasionally the pipe may be enclosed in a buttress on the side of a wall as in VI.1.

Upper down pipe fitting tightly into the one below

In most cases the pipe fits closely within the masonry of the wall and is only visible because some of the outer covering, be it plaster or stonework, has fallen away as in VI.4.1,2 and VI.6.9. Masonry work can be a simple fill across a corner as seen in VII.4.38 or square as in VII.3.35 and VII.9.50. The placement of these pipes with the properties in most cases suggests that they are emptying into cess pits. This scenario particularly applies when the pipes are associated with an outside wall of the property (e.g.VI.34.1,2: VI.6.9). In a number of cases the latrine on the ground floor, within the house has a down pipe, from an upper storey latrine, associated with it.

The pipes are of different diameters, varying from the largest of 25cm, to a small size of 10cm. Narrow bore pipes may be the ones which brought water from roofs into cisterns. One of these has definitely been identified associated with the cistern in VI.1.18,20 (AAPP unpublished excavation report). Others may be as illustrated in I.4.9 and in the 2007 excavation by a Spanish group in VII.4.30 where a down pipe appears to be feeding a cistern or tank. The down pipe in the outer wall of IX.1.28 appears to empty onto the sidewalk and there is a groove in the kerbstone worn by the water from the roofs above. It is of prime importance to date all these pipes. This would aid considerably in archaeological phasing.

The archaeological evidence of the positioning of cesspits has thrown some light onto a number of questions. Jansen has suggested that these are positioned outside houses so that ‘the pit could be emptied without causing a mess in the house’. In addition, and somewhat contrarily, she also suggests that the cesspits in Pompeii would not require emptying because liquids could permeate ‘through the seams between the bricks’ (Jansen 2000a: 38). There are undoubtedly other reasons for having cesspits in the ‘open air’ of the street (though they almost certainly had covers of wood or stone). These include the obvious physical (e.g. the obnoxious smell), the cultural (e.g. attitudes to excreta), and the chemical (e.g. the production of methane from the waste). This latter product might have provided a known danger to someone using the facility holding a naked flame in a lamp, although, surprisingly we do not appear to have any reference to explosions in latrines. In the cases where the excreta drain from the latrine into the street the base of the latrine is itself sloping and tiled or made of cement (See VI.11.6,13). There are some cases where the cesspit appears to be inaccessible because the street and the sidewalk are completely paved.

In order to draw conclusions from the evidence which has been collected it is necessary to consider the development of latrines to see if there were any changes in the technology or in the cultural attitudes. These will include concepts of privacy, gender issues, and the local attitudes to rubbish disposal and pollution and health.

Before these are explored the bibliographic work on this subject should be stated and it is to this that the next chapter is devoted.

PREVIOUS BIBLIOGRAPHY

Eighteenth and nineteenth century archaeologists were particularly concerned with the forum, the tombs and the theatres etc. and had little interest in toilets - even when they were describing the houses (Breton 1855, Horne 1895). De Jorio, an archaeologist and a priest, was either confused about the two multi seat latrines in his description of the excavation of the *macellum* in Pozzuoli, where he described them as being bath houses, or perhaps his sensibilities were offended De Jorio 1820) . An exception to this was Gusman who at least shows a drawing of a latrine but unfortunately neither gives us its location nor expands upon it in any way (Gusman c1900: 282).

The *Dictionnaire des Antiquites Grecques et Romaines* (Daremberg and Saglio 1873-1910: 987-91) gives an extensive entry on the subject of the 'LATRINA'. This deals in some depth, and with a high degree of accuracy, with the latrines in Pompeii commenting that: 'without exception every house in Pompeii has a latrine'. It describes their situations: 'The custom of placing the latrine in the kitchen is far from being the general rule' and states 'Often the latrine is to be found in the depths of the house, or else opening onto the hallway or using spare space under the staircase'. It also mentions those of the upper storeys: 'On the first storey there were latrines which connected by a pipe with those of the ground floor or were provided with their individual waste pipe'. Much of this entry on latrines, which was written by Henry Thédenat, has been confirmed by examination of latrines throughout the city and in particular the illustration in the *Dictionaire* of the latrine and a down pipe is easily recognisable as the one in VII.9.63.

Fiorelli was certainly fully aware of the latrines within the city and in the Appendix to this volume all of the latrines he records are mentioned (Pappalardo, U. 2001. [Fiorelli 1875]). Sogliano recorded, in the Notizie Scavi, the excavation of four latrines (Sogliano 1900). These all drained into cesspits of varying depths and constructions. They were as follows. The first at VI.10.14 was 11.2m deep with a circular diameter of 1m. The one at VII.14.25, of rectangular construction 1.5m x 2.2m was 6.3m deep. The third in a shop, VII.12.11, was an ellipse with the greatest axis of 1.5m and smallest 0.95m and the total depth was 5.40m. Finally the fourth latrine described at VII.7.10

resembled a cistern, lined with plaster, with a mouth of diameter 0.60m and a bell-shaped expansion below the ground to a depth of 5.3m. This 'cesspit' also had a branch at the bottom turning south towards the Via Marina. The contents of three of these cesspits were of 'volcanic debris' (probably a reference to *lapilli*) and only in VII.14.25 was discovered 'one dark coloured stratum from whence came a nauseating smell'. Currently there are problems with the excavation to these depths because of Health and Safety regulations.

Ten years after Sogliano's publication, Thédenat discussed latrines and their positioning within the Pompeian dwelling. He noted their initial proximity to the cooking area and also that, over a period of time, they were moved to other parts of the dwelling and to the upper storey. He commented that rarely were the latrines connected to sewers, but unfortunately gave no examples. He did, however, give five examples of where latrines emptied directly onto the road or pavement [II, 4, 41; III, 7, 1; IV, 3, 24; IV, 7, 10; VI, 15,5] (Thédenat 1910: 96). Each of these addresses has been checked and no evidence has been found of latrines emptying directly onto the sidewalk. Perhaps Thédenat was implying drainage of latrines into cesspits in the sidewalks. These occur all over the city of Pompeii, not just as described in V1.1. However, there is no evidence of these features either, although they could be lying, invisible, beneath modern surfaces.

Occasionally early writing related to water drainage systems. Mau writing at the end of the 19th century makes an accurate interpretation of what occurs in Pompeii. He notes that; 'There were covered conduits to carry off the surface drainage from the forum, one of which runs under the Strada delle Scuole to the south, the other under the Via Marina to the west. Elsewhere the water rushed down the streets till it came near the city walls, where it was collected and carried off by large storm sewers. One is at the west end of the Vico dei Soprastanti, another at the west end of Nola Street, and a third leads from Abbondanza Street, where it is crossed by Stabian Street, toward the south'. He goes on to say that 'there were other sewers in the city, but they were of small dimensions and had not been fully investigated. In general they seem to have been under sidewalks. They were not designed to receive surface water, but the drainage of houses. They cannot have served this purpose fully, however, for *most of the closets were connected, not with the sewers, but with cesspools.'* (Mau 1902:229).

Early in the 20th century a work was published entitled *Hygienische verhaltnisse im alten Pompeji (*Mygind 1921). The author, Mygind, describes the different situations within the properties in which latrines are found - noting that each house had its own, even the smallest dwelling. Latrines, he states, were usually sited somewhere secluded in the house. If they were in a separate room it was usually a very small one without a door and possibly with a curtain across the opening. He notes that, if the latrine was in the kitchen area, there was sometimes a chest-high wall separating it from the cooking surface. He believes that ash was used to cover excrement in the cesspit to reduce smell, or perhaps to create garden fertiliser, finally he comments that some kitchens had window openings and holes in the ceiling to allow fumes to escape. He goes on to state that many shops, workshops and most drinking establishments also had latrines. He suggests that some latrines were two seaters (IX. iii. 31) or even three (IX. ii. 27). He affirms that no waste pipes from toilets open into the street and observes that there were latrines on the first floor with visible pipes down to cesspits, giving as examples I.1.8: VII.2.20: I.2.7: and VIII.4.4. He describes sloping platforms making it easy to keep the toilets clean and suggests that they could have been used for the disposal of waste water. Perhaps mistakenly, he surmises that the seats were probably made of stone and often just rested on plinths because the slot was much too wide for even the heaviest wooden plank. He notes that a stone seat was found at IX.5.9 but 'is no longer there'. He also comments that all the public toilets emptied into sewers whereas only a few latrines in private houses did so. He construes that *graffiti* indicate that people defiled the streets and public places with faeces. Interestingly, he states that latrines were used by family members, slaves and clients but then goes on to say that the upper classes used chamber pots or portable commodes. Mygind's work, which although published over nearly ninety years ago still stands out as a masterpiece, but unfortunately has never been translated into English, which is a great loss to students who do not read German.

Unfortunately little more came of the subject of latrines until the publication of a wide-reaching article on 'Slums, Sanitation and Mortality in the Ancient World' (Scobie 1986). Linking the sanitation of Roman cities to perceptions of life expectancy and mortality in the Roman Empire, this broadened the perspective of the whole picture by exploring the physical aspects of latrines and effluent disposal. Within this work, areas for research were identified which subsequent workers have devoted much of their time to answering. This work by Scobie has been expanded by Andrew Wilson who has given an extremely succinct, historical and descriptive, analysis of drainage and sanitation systems in the ancient world (Wilson 2000).

In the last decade of the 20th century most of the new work has been attributable to Gemma Jansen (Jansen 1992, 1993, 1994, 1997, 2000a, 2000b, 2001), with some work by Richard Neudecker (1994) Ann Koloski-Ostrow (Koloski-Ostrow 2000, 2001) and Natalie de Haan (de Haan 2001).

Examination of single properties within Pompeii has resulted in publication of many outstanding reports among which Descoeudres & Sear (1987) and Ling (1997) are to be reccommended and also the magnificent series of descriptions of Pompeiian houses 'Haüser in Pompeji'(Ehrhardt, W. 1988: Michel, D. 1990: Seiler, F. 1992: Stemmer, K. 1992: Strocka, M. 1984: Strocka, M. 1991: Fröhlich, T. 1996: Ehrhardt, W. 1998: Allison,P., Sear, F. 2000). Much may be gained by referring to these publications but in contrast this volume offers a wide based catalogue of all the latrine systems of the city.

UPPER STOREY LATRINES

The destruction of the upper storeys of Pompeii has left many questions unanswered about the usage of the upper floors. Upper storey latrines were commented upon in an article discussing private latrines in Pompeii (Jansen 1997, 125), but there has been little further research over the past twelve years and more information is required about these and about their terracotta down pipes, which are a common finding throughout the city. Throughout the city the authors have identified nearly 200 down pipes. In one case the name of the pipe maker can be seen, *C. Juli(us) Nonors*. The upper ends of the vast majority of these pies, running in the walls of the city, are broken and give no hint of where they are from. Their existence is said to confirm upper floors (Pirson 1999, 170). The pipes can be seen in different parts of the properties e.g latrine rooms and peristyles, and they are of varying diameters and different materials. Two main functions can be postulated for these pipes. Firstly, they may transmit water from roofs, for drinking or other purposes, or they may be carrying effluent material from upper storey latrines. In order to distinguish between these two functions a number of factors relating to the pipes must be studied. Laboratory analysis of the contents of some of these confirms that they carried excrement from latrines (Love 2007). However, other than the identification of mineralised faecal material from within the pipes are there any other features which may help to differentiate the two functions? Most of the pipes have flanges which secures a tight fit from the upper pipe into the one beneath it. This type of construction will only allow the passage of material downwards and will prevent seepage from the junction. If water is to be transmitted this might not be too important, but excrement must be contained. However other building features may also have played a part in reducing problems such as leakage or smell which might be associated with a pipe which is transmitting faeces and urine. In many cases the pipe fits closely within the masonry of the wall and is only visible because some of the outer covering, be it plaster or stonework, has fallen away as in VI.4.1,2, VI.6.9, VII.13.25 and VIII.7.5. Masonry work can be a simple fill across a corner as seen in VII.4.38 or square as in VII.3.35 and VII.9.50 or occasionally the pipe may be enclosed in a buttress on the side of a wall, as in VI.1.1

The diameters of these pipes vary from the largest of 25cm to a small size of 10cm. A narrow bore pipe, transmitting excrement might easily become blocked especially during hot periods of the year, despite water being poured down them and it is suggested that narrow bore pipes carried water. The placement, within the properties, of the large diameter pipes, in most cases suggests that they are emptying into cess pits. This scenario particularly applies when the pipes are associated with an outside wall of the property. In a considerable number of cases the latrine on the ground floor, within the house has a down pipe associated with it, emptying into the same pit. If the large diameter pipes are traced upwards a number of upper storey latrines have been defined (22 in 2009). In the case of the upper storey latrines in VII.12.14 and VIII.4.37 the presence of two down pipes has been commented upon suggesting latrines not only in second storeys but perhaps third storeys also and IX.2.7,8 has a pipe from above going into the down pipe from the first storey latrine. This may carry water for flushing. At ground floor level double down pipes can be seen at IX.1.9 and IX.3.15. It is obvious that these are coming from separate sources, and may also indicate second and third storey latrines

Narrow bore pipes carrying water may be collecting this from roofs and transmitting it into cisterns. One of these has definitely been identified associated with the cistern in VI.1.18, 20, and there are a number of others as in I.4.9 Not all of water pipes go into cisterns and the down pipe in the outer wall of IX.1.24 appears to empty onto the sidewalk and there is a groove in the kerbstone worn by the water from the roofs above.

It would be of prime importance to date these pipes. Archaeological evidence and phasing is difficult but as an example the masonry of one of the enclosed pipes is composed of *opus reticulatum* (VI.14.30,31,32). This is suggestive of first century CE construction technique.

All of this technology must have related to cultural attitudes relating to smell and waste disposal and these are now to be discussed.

DIRT AND SMELL AND RUBBISH DISPOSAL IN LATRINES IN POMPEII

The concept of 'dirt' in any community varies according to a number of different factors and may have changed over time. Pollution by 'dirt' and/or 'smell' evokes degrees of response from individual horror and disgust to social and gender discrimination. Body emissions, including human excrement are well documented, in modern terms, as highly defiling (Clark and Davis 1989). Was it different in Roman times and do the latrines in Pompeii give us any information?

It is reasonable to assume that the latrines in Pompeii were malodorous. Scent travels more effectively in hot, humid air (Bartosiewicz 2003:179). Many of the latrine rooms in Pompeii have windows. These may have provided a degree of ventilation as well as some degree of light. There is, however, some evidence that there was glass in some of the windows (Hilary Cool: Pers. Com). What alternative might there have been to control the smell? Emptying burning charcoal from the kitchen into the toilets would not only have sealed the contents but also would have burnt off the methane, which is produced by the decomposition of the excreta. Excavation of cess pits has provided little evidence, as yet, of this practice. There is no doubt that the characteristic smell of faeces, provided by the chemicals Indol (C8H7N) and Methylindol (a.k.a. skatole: C9H9N), must have lingered in the air. The decomposition of urine must have contributed to this with the smell of ammonia.

Throughout the city of Pompeii there were shops and bars selling food some of which was being cooked, and there also were a number of fulleries and dyehouses from which distinctive aromas would permeate the streets. The odour from cesspits may have been a further problem in the streets, despite that they were probably covered by stones or by wooden covers. These may have helped to contain the smell. However, hot climatic conditions with high humidity and low air circulation, particularly in the summer months are likely have increased the smell prevalent in the streets (Dague 1972: 592).

There is some evidence that flowers or herbs were hung on the walls of the latrines. Perhaps that was partly to disguise the smell.

In some houses a tiled surface may not necessarily signify a toilet. In the corner of one of the houses in the southern part of Pompeii I.17.3 is a quarter of a circle feature with a drain leading out - presumably into a cesspit. There is no evidence that this is a toilet and an assumption is made that it is for rubbish disposal, possibly of a 'watery' nature. Similar chutes occur elsewhere, one being at floor level alongside the kitchen working surface in VIII.2.24. This type of liquid rubbish disposal gutter was called *trua* (Scobie 1986. Footnote 88:410).

VIII.2.24 ?Waste chute

I.17.3 ?Waste disposal

There are drain holes in walls near latrines which exit into the street (e.g. VII.6.30,37) and in the House of the Vestals a terracotta pipe runs through a wall into the latrine cess pit from a room in the kitchen area where ecofact evidence of food preparation was found on excavation.

VII. 6. 30, 37 After cleaning:

An interesting feature can be seen in VIII.4.53 which may be for rubbish disposal but more probably is a cistern head. Evidence from excavation is required.

VIII.4.53 ?for rubbish disposal

VIII.4.53 Feature from above

PRIVACY

The concept of privacy in Roman latrines merits considerable discussion (See Hobson 2009). There are gender issues as well as social and personal factors. The types of clothing, particularly long tunics worn by women, may have aided in situations where both genders may have been using multi seat latrines at the same time.

There is considerable evidence for privacy in the latrines in Pompeii. Even in the kitchen areas, where the latrines appear to be related to the lower classes, there is evidence for wooden partitions which separated the latrines from the cooking surfaces (e.g. VI.2.16, 2). In some of the working areas of large houses (e.g. House of the Vestals VI.1.6,7,8,25,26) there is a room some distance away from the work areas. In the kitchen of VI.5.9,19 there is a definite room within the kitchen area with grooves in the threshold stone where door posts would have fitted. Throughout the city there are considerable numbers of thresholds leading into latrine rooms with similar slots for door posts.

The latrines adjacent to the streets and by entrances to the properties appear to have had a considerable degree of privacy and those to the rear of workshops and bars were similarly protected from 'public' gaze.

Despite this some latrines are apparently not so private. An example of this is the latrine in the kitchen of VI.2.27. The upper storey latrines are in niches set into the outer walls. There is no privacy for the users of these toilets. However, since they are almost certainly family orientated it may be that privacy was not an issue.

THE USE OF WATER: DRAINS AND SEWERS

The Romans were fully aware of the importance of water in their lives. Celsus differentiated a range of quality, from rainwater, which he called the lightest, to marsh water, the heaviest (Celsus II.18.12). Pliny the Elder also discussed quality, although he found difficulty in accepting rain water or snow as being the best because he recognised the presence of pollutants from the air in both (Pliny XXX1.xxi/xxii).

Early in the urbanisation of cities it was recognised that water, in quantity, was essential. Rainfall, springs, rivers and later, aqueducts, provided this necessity. Surpluses, such as those occurring with heavy rainfall upon hard surfaces, for example roofs or streets, would require either storage or disposal, and the unwanted excess required removal or drainage systems.

Household water could be obtained by digging wells down to the water table or by tapping springs. In Pompeii wells were dug through the lava spur upon which the city stands. Most of these appear to have been situated in public places, providing water for everybody, although some houses had wells of their own. Rainwater was collected off roofs and was collected into cisterns. These provided storage for periods of time when rainfall was scarce (Vitruvius VI.14). Usually collection was from the *compluviu, the* water falling into the *impluvium* but alternatively narrow bore pipes might bring the rainwater directly off the roof into the cistern.

Cisterns abound in Pompeii. At some time around the fourth century BCE, waterproof cement lining was invented which prevented seepage out from, and contamination into, the cisterns. These varied in size. For example the one in the House of the Painted Capitals measures 12.5m by 2.2m and was 2.7m high - and would have held about 69,000 litres of water (Sear 1994). During periods of heavy rainfall the cisterns would fill and overflow; the water then ran out through drains under the door thresholds into the streets, helping to cleanse them.

Mau discusses the surface drainage system of the city of Pompeii, noting that “there were covered conduits to carry the surface water from the forum” and that these ran “under the Strada del Scuole to the south and the Via Marina to the west”. He also states that there were exit drains at the west ends of the Vico dei Soprastanti and the Strada di Nola. His final comment involves a drain which he says "leads from the Abbondanza Street where it is crossed by Stabian Street toward the south" (Mau 1902: 229).

In his publication of the work carried out in Pompeii in November 1900 Sogliano was more precise (Sogliano 1900). He explains that the flowing of the water from the upper part of the city to the lower occurred given the sloping nature of the terrain and that as well as rain water the overflows from the fountains and from the houses contributed to the flow. Because of this he comments that "from time to time, in appropriate places it was conveyed in underground drains that lead out of the built up areas". Here he particularly mentions a junction between drains uniting at the western junction of VI.4 and the Via Nola (Via delle Terme) and then going south taking water from VI.1, V1.2, VI.3, and VI.4.

Although he is not totally explicit he says that another drain runs under the northern aspect of VII.15. The full extent of this was not known at the time since the buildings of the western insula had not been excavated. He then goes on to say that the forum had drains on the east, west and south sides leaving down the Via Marina between the Basilica and the Temple of Apollo. He further notes that other "minor courses" joined this (but were only partly excavated). Most interestingly he describes a huge cistern under the south side of the forum approximately fifty metres long by two and a half metres wide. He says that the water flowing into this would be less impure than that from other areas since "the forum was inaccessible to animal drawn vehicles". A separate series of drains he describes running under the Via Abbondanza, one part passing south under the Via Stabiae to leave the city, the other apparently going under the large theatre towards the north west corner of the Ludo Gladiatorio and finally passing out through the city wall. Again he notes that "a number of water courses and foul waste water drains join this", some of which "appear to have been made after the original construction – the intention of which, at first was the single purpose of collecting rainwater". It is important to note that at no time in this paper does Sogliano suggest that latrines are discharging into these drains.

Further work by Paul Arthur (Arthur 1986: 37) expands the understanding of the drainage in the forum area. This work states that during the late Republic the forum was equipped with a central sewer with collateral culverts. In Imperial times run-off from the forum was directed through a series of holes cut in the travertine blocks lining the square into a number of conduits which converged with the main sewer to sluice beneath the Via dell' Abbondanza or the Via Marina." He regrets, "for safety reasons, being unable to draw up a plan of the entire system" but states that there was "a sewer draining the Temple of Apollo and another, now sealed off, which commenced in front of the House of Triptolemus". He also found evidence of earlier, minor, drainage systems using amphorae.

Alongside, and yet inextricably mixed with this, was the problem of the disposal of rubbish, and human effluent. As Wilson says, "Storm-water drains are necessary if these impermeable surfaces are not to flood, and such drains *may also serve as sewers.*" (Wilson 2000: 151) {authors italics}.

Previous comment has been made about sewers from public toilets. That there were such is confirmed by Eschebach in his description of the Stabian Baths (Eschebach 1979). Not only does he describe the sewer as starting from the centre of the rear seating row but shows it running south towards the Palaestra and then being joined by other drains from the baths, eventually going into the main drain in the Via Abbondanza.

A desire to imitate Rome in its usage of water for public and private display (fountains, and fish and swimming pools etc.) led to the construction of an aqueduct to Pompeii at the time of the early Empire. This aqueduct provided a major change in the supply of water, bringing it to a *castellum* at the highest point of the city from where it was distributed, using lead pipes, to public fountains, workplaces and private houses. This was carried out via a complex tower and siphon system with taps to control supply. The piped water gave new opportunities for aquatic display within houses and the mere sight of this expensive luxury, which for private consumption had to be paid for, pouring out through the overflow drains, would signify the elite status of the householder and his apparent disregard for the expense. Despite this apparent accessibility to piped water, there is little evidence to suggest that it was brought to kitchens for drinking and cooking.

Rome, also on a sloping site, had a similar type of street cleaning. Frontinus comments about overflow water from reservoirs and fountains.

> *Nam necesse est ex castellis aliquam partem aquae effluere, cum hoc pertineat non solum ad urbis nostrae salubritatem, sed etiam ad utilitatem cloacarum abluendarum* (Frontinus: Aqueducts II.111)

For there must necessarily be some overflow from the reservoirs, this being proper not only for the health of our city, but also for use in the flushing of the sewers.

Tiled floor in VII.15.4

What evidence is there for the use of water to flush the toilets? Firstly there is the presence of sloping tiled floors. Many of the toilets in Pompeii have sloping tiled floors. This would appear to be confirmatory evidence that water was used as a cleansing agent. However, perhaps the sloping tiles were there purely to direct urine if males were micturating standing in front of the toilet seat. This concept presupposes that Roman males urinated whilst standing up!

Next there is the presence (or in VI.1 the absence) of water tanks. It has been said that private toilets usually had some sort of water basin (Jansen 2002: 60). None of the systems in VI.1 have such a provision. In fact, few toilets in Pompeii have associated water tanks, a rarity being seen in 1.11.11 (See 'Decoration in Latrines'). This is a latrine which has its entry from the garden. It is probably a late addition to the house facilities and may have originally been a woodshed.

Two other latrines with water tanks are worth noting, both in working areas. They are the latrine next to the kitchen of the Praedia Julia Felix and the latrine in the kitchen of the house of Caecilius Iucundus (V.1.23,25,26,10). The water within these tanks might have been available for usage in the kitchen as well as for the latrine.

In the majority of cases the water would have been brought to the latrine in buckets. Repetitive usage of a latrine in a *taberna,* for example, might require regular flushing with the water from a bowl or bucket and perhaps its usage as a sponge cleanser. That process would necessarily pollute the water, which might occasion it to be renewed regularly with buckets from an outside source, the most likely being the local street fountain.

Niche toilets in upper storeys would have required flushing to keep the pipes clear (Jansen 1997:124ff). Most modern authors have assumed that flushing with water did take place and that it was primarily for hygienic purposes. As has already been indicated there is no doubt that, in addition to this function, the bacterial decomposition of the excreta in the cess pits would have been accelerated by the persistent addition of water, especially in the warm climatic conditions of the city of Pompeii during the summer months.

VII.13.14 Latrine showing water channel running across the floor in front of the toilet

Piped water is, in fact, supplied to a small number of toilets. "…. two toilets had running water; in the toilet of the Casa del Nozze d'Argento, a tap is attached in the wall: and in the Casa del Fauno, a hole for a pipe can be seen next to the toilet seat and a groove for a pipe cut out in the outer wall of the house" (Jansen 2001:40 Note 42).

Another author states "I have thus far found no evidence for piped water into a kitchen or latrine, except in the latrines associated with the baths in the House of the Centenary and in the House of Julia Felix" (Jashemski 1979:53). In making these comments there is no clarification about the actual use of the water for flushing the latrine. At this point the question must be asked: If water was worth piping to one or two latrines and so much of it was being put down latrines then why are there not more water pipes in or near latrines? Cheap labour could have brought water into houses from the street fountains but installation of a lead pipe supply would have been expensive. This suggests that the use of running water was a social behaviour relating to the rich sections of the Roman society?

VII.13.14 Water Pipe emerging through wall and groove of kerbstone

To clarify this a little there is a latrine in the House of Ganymede (VII.13.14) which has water running across the room, in front of the latrine, to exit through the wall to the north via a terracotta pipe (see illustration).The southern end of this pipe abuts the southern wall of the room. The northern end of the pipe runs out through the north wall and its contents would have passed down the groove in the kerbstone and dropped onto the road surface from where it percolated through into the cess pit below the road. This pipe carrying water across the room in

VII.13.14 Lead pipe going into wall from south to deliver water to latrine

front of the toilet could easily have been diverted into the toilet.

Lead piping running down street diagonally opposite to VII.13.14

On the other side of the wall is a lead pipe entering the wall. This appears to be **ascending from the floor below,** though there is no positive evidence for where it was coming from. Piped water is, in fact, entering the house only a few metres away. The lead pipe crosses the street and ascends up a groove in the kerbstone to enter the house at its north-east corner. The illustration shows the lead piping in the street diagonally across from VII.13.14. This is coming from the fountain supply up the street to the north. It would appear that, from the Augustan period when piped water was available for display in the fountains of the elite houses, this particular house used this water supply system to provide intermittent running water through the latrine room. Since it was running directly onto the street it would seem likely that cleansing of the toilet itself was considered unnecessary but that the tiles could have been washed from the supply.

VII.13.14 Groove in kerbstone for lead pipe coming from street and going through wall into house

Finally, in attempting to demonstrate the use of water for

flushing, there is evidence that the waste water of the *caldarium* of the Casa del Fauno was used to flush the latrine in a room south of the *caldarium* (De Haan 2001:44) and also that the water from the pool of the Praedia Felicis was relayed through the multi-seat latrine to wash the effluent into a cesspit (Parslow 2000, 201 and plan 203). In the Casa del Granduca VII.4.56 the toilet is situated beyond the kitchen at the southern end of the property. A pipe runs from the gutter of the peristyle through the kitchen and into the toilet and it has been suggested that this is a way of disposing of the water from the fountain in the peristyle (Descoeudres & Sear 1987:30. Sear 2004:162). Similarly the latrine in room 26 of the Casa del Menandro was flushed by water from a sink in the kitchen (room 27), to the north (Ling 1997:92f.). It is also thought that down pipes carried rainwater to flush latrines in both I.10.2,3 and the Casa del Fabbro I.10.7 (Ling 1997:41 and 145f). Unfortunately there is insufficient information about the diameter of these down pipes and they may have serviced latrines in the upper storey.

Mau shows a drain from the large swimming tank in the Central Baths passing to the public latrine, presumably to act as a sluicing system. However, he does not indicate the sewer drainage from that latrine (Mau 1902:209). Although no excavation had been carried out to confirm this, it is believed that the drainage from that latrine is into a large cesspit (De Haan: 2004 pers. Comm.)

The earthquake of 62CE caused major damage to the city of Pompeii. One of the facilities affected was the water supply, and recent excavations in the Casa delle Vestali shows that the house underwent major changes to enable its owners to cope with this loss. It is possible that some latrines were modified so that less water was being used to flush the tiles. New research is required to confirm this.

DECORATION

Inevitably there will be discussion about the decoration of rooms which may be considered to be among the least important in the properties. Visible signs of decoration, in the latrines of Pompeii, are minimal, due to the deterioration of the buildings over the past two hundred and fifty years. Disappointingly the walls of one of the most decorated latrines, within the House of the Silver Wedding (V.2.i), have now collapsed, and will not be repaired. In other latrines decorated plaster has faded even in the rooms where it still exists. This lack of decoration makes it difficult to assess the significance of different styles.

Latrine in House of Castor and Pollux: VI.9.6

Eight decorated latrines in Pompeii are mentioned in the article 'Paintings in Roman Toilets' (Jansen 1993.33 Note 2). The list of includes I.11.10 the garden toilet of the House of Euxinus, Two toilets in II.4.3 the Praedia of Julia Felix, V.1.18 the House of the Greek Epigrams, V.1.30, V.2.1the House of the Silver Wedding, VI.9.6 the House of Castor and Pollux, and the public toilets of the Stabian Baths and the Suburban Baths.

Floor of House of Castor and Pollux

Of these the House of Castor and Pollux serves as the best example. This latrine has a high decorated dado, painted red, above which a garland of flowers is visible. There are yellow panel lines but no other motifs. The floor is sloping and is made up of tiles in a variety of fish scale construction.

The quality of decoration can be seen in the House of Jason (IX.5.18) where, though only a small amount of wall plaster remains, the colours are striking. The lower orange is separated from the upper red by dark red lines.

IX.5.18 House of Jason

The latrine of the garden of I.11.10 has a red line separating the black dado from the paler area above. Jansen states that this is a general Roman tradition since it has been noted in latrines at Ostia (Jansen 1993.30)

I.11.11 Water basin

The multi seat latrine in the Praedia Julia Felix has a yellow dolphin on the north wall.

As mentioned earlier the latrine in the House of the Silver Wedding is no longer viewable. The room was vaulted (see also VI.7.20,21,22) with a small round window. There were sprigs of red flowers on the walls and the ceiling had a seahorse in the centre with dolphins and drinking horns (Jansen 2003.32). On the wall to the right of the entrance to the multi-seat latrine in the Suburban Baths there is a good picture of the goddess Fortuna.

Fortuna on the wall of the Suburban Baths

This goddess appeared on the walls of IX.7.21,24, also of V.1.18 (no longer visible) and of II.4.3 (Archivo Fotografico 8566 neg 1404) (Jansen 1993.33 Note 5) and there has been considerable discussion about the role of the gods in the process of human excretion.

Graffiti associated with different aspects of life at the time ranging from electioneering to coarse comments on the quality of food, wine or sex have been recorded from all parts of the Roman world. In Pompeii alone there have been fifteen thousand recorded. Gods and goddesses are particularly associated with graffiti found in or related to latrines. Those associated with urination are less common than those relating to defaecation. An inscription on the forum arch at Thigibba in Tunisia reads:

"si qui hic urinam fecerit, habebat Martem iratum" (Quoted in Wilson 2000)

Anybody urinating here will incur the wrath of Mars.

Food, wine and sex all produce comments, often in or around work places such as bars or the brothel at VII.12.18-20. One man, Aemilius Celer has been identified as the 'artist' associated with at least thirty five (See Franklin 1991: Footnote 40). There are six examples of alphabets scratched, perhaps by children, low down on the walls. Writing on latrine walls appears to have been as common then as it is now. Obviously this indicates some degree of literacy. However, in the Pompeiian graffiti a high percentage have spelling mistakes after the first two or three words and this may lead to the assumption that the people who wrote the majority of these comments were not of the elite class.

In Pompeii there was a graffito to the left of the door of the House of Pascius Hermes which is virtually the same, only the god is different.

"Cacator cave malum aut si contempseris habeas Iove iratum" (CIL 4:7716)

To the one defaecating here beware of the curse. If you look down on this curse you will have an angry Jupiter.

By the sepulchre of Caecilius Felix was the following.

"Qui hic mixerit aut cacarit, habeat deos superos et inferos". (CIL VI:13740 Trans. Pascal)

May he who pisses or shits here incur the wrath of the gods above and below.

On a wall of IX.7.22 and also a similarly in V.15 in Pompeii could be found

"Cacator cave malum"

When you relieve yourself beware of evil.(CIL 4:3782. CIL 4:4586 Cave malum cacator)

Also "*Cacator cave malu*" was to be found, on a picture of a man between two serpents and the goddess Fortuna, by an entrance to a latrine in Regio IX. (CIL 4:3832). Pompeii has other warnings about relieving oneself inappropriately. In Regio V was written

"*Stercorari ad murum progredere si pre(n)sus fueris poena(m) patiare neces(s)e est, cave*". (CIL 4:7038)

If you shit against the walls and we catch you, you will be punished.

The wall opposite the latrine in the House of the Centenary (IX.8.3,6) in Pompeii was full of graffiti. A modern poster on this wall describes three of these.

"Quodam quidem testis eris quid senserim quando dovro cacare, varro qui a cacare"

You will be the witness of what I have experienced. When I have to poop I have to poop

"Marthae hoc triclinium est, nam intriclinio cacat"

This is Martha's triclinium: in fact she poops in the triclinium.

It seems that one graffito was not enough for some people because on the same wall appears three times the following remark.

Secundus defaecated here. (CIL 4:5243)

Perhaps somebody doesn't like someone else in the house because on the same wall is written:

Quodam quisem testis eris quid senserim (νδι) cacatuiero veniam cacatum (CIL 4:5242 Trans. Pascal)

Someday indeed you will learn how I feel. When you begin to shit I will shit on you

Comments about defaecation do not always appear within the confines of the latrines. Just inside the Vesuvian gate in Regio V was this one.

"Cacator sic valeas, ut tu hoc locum transeas" (CIL4:6641)

Shit with comfort and good cheer, so long as you do not do it here.

This is obviously a warning but is it purely to prevent a dirty mess or are there overtones of religion or health. Somebody may have a weird sense of humour! This appears on a wall in Regio II.

Lesbianus shits and writes 'Hello'. (CIL 4:Supp 3. 10070)

APPENDIX

APPENDIX

The following pages are primarily a photographic record of the latrines in Pompeii. Fiorelli instigated the division of the city into Regions and Insulae, and allocated individual numbers to entrances in each insula. This method of identification has been followed by subsequent archaeologists with a small amendment in that a few entrances are denoted by letters. Occasionally a blocked doorway has either been given a number or completely ignored resulting in some confusion in the record. Room numbers are copied from Bragantini, I.1983. Pitture e Pavimenti di Pompei.

Down pipes are indicated by the symbol ⊕ and Latrines by the  figure

There is no plan of Regio IV. This is because it has no properties which have been fully excavated. There are also some insulae in other Regions which are incompletely excavated and are only identifiable by the fronts of properties which line the streets. Such are I.18. and I.19. No plans for these are included. Plans for one or two insulae are inaccurate and an attempt has been made to indicate where the latrines and/or down pipes exist in these.

The photography has been made difficult due to the existence of shadowing in the bright sunny conditions which exist in the months of July and August.

Apologies must be made for any errors, particularly if the situation of a latrine is wrong.

REGIO I

REGIO I. 1

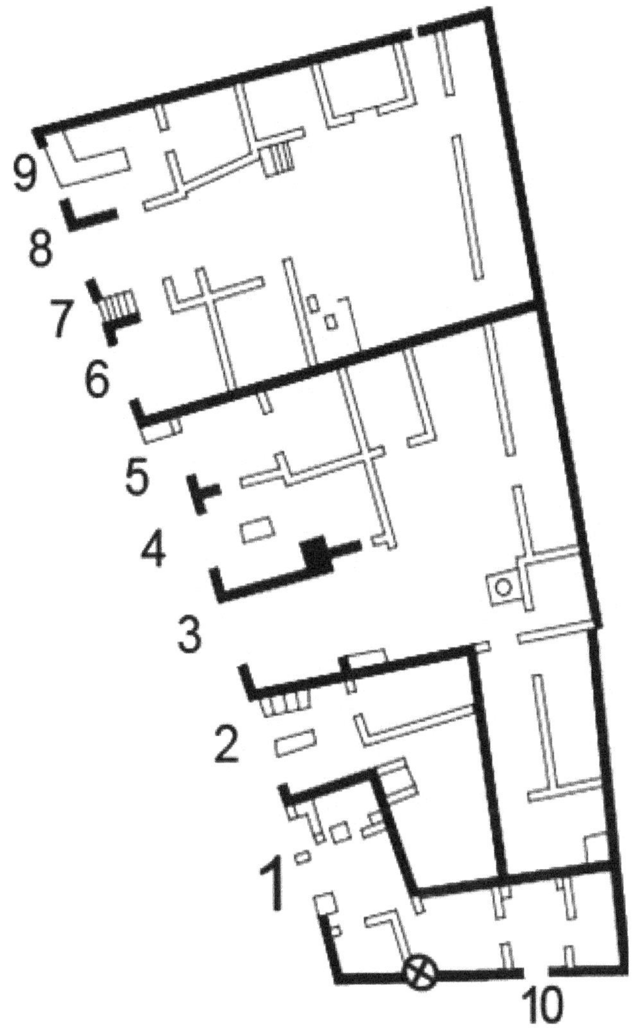

A small insula with only four properties.

In 2006 the southern properties I.1.1,10 and I.1.2 were totally overgrown with vegetation. No latrines or down pipes found because of the dense undergrowth. In 2007 most of this had been removed. Despite this none of the previously recorded latrines were identifiable. One down pipe, enclosed in masonry, was observed in I.1,10

I.1. 1, 1a, 10 ***Caupona des Espegatus Cilo***

Latrine in 1a (Eschebach p13)

DOWN PIPE in masonry buttress against south wall

I. 1. 2 ***Caupona***

Latrine by kitchen (Eschebach p13): (Fiorelli p33) No evidence

I. 1. 3, 4, 5 ***Stabulum***

'Spaziosa' latrine in 3 (Fiorelli p33) No evidence

I. 1. 6, 7, 8, 9 ***Stabulum des Hermes***

Latrine in 8 (Eschebach p14): (Fiorelli p33): (Packer p6) No evidence

I.1.10 Down pipe

REGIO I. 2

This insula has a total of fifteen properties most of which are small houses. In 2006 the houses with numbers from I.2.15 to I.2.29 were full of undergrowth. Most of this had been cleared by 2007. Disappointingly in none of the properties are there any latrines visible despite descriptions by Eschebach Fiorelli and Packer. Three down pipes have been found corner of #8 room b.

I. 2. 1, 30, 31, 32 Popina

Nothing

Latrine in 1 by entrance #30 'under wooden stairs' (Eschebach p14: Fiorelli p34)

I. 2. 2, 3, 4 Casa con colonna Etrusca

Nothing

I.2.5 Shop

Nothing

I.2.6 Casa degli Attori

Nothing: Latrine (Fiorelli p35)

I.2.7,8

DOWN PIPE in 8 room b

I.2.9.10 Atriumhouse of L & M Volusius Faustus

Nothing

Latrine in 10 (Eschebach p16: Fiorelli p35)

I.2.11 Shop and Thermpolium

Nothing

I.2.15 House

Nothing

I.2.16 House

Nothing

I.2.17,18,19 Brothel

Nothing

Latrine in19 (Eschebach p17: Fiorelli p36)

I.2.20,21 Caupona, thermopolium & brothel of Pollius

Nothing

I.2.22 Shop

Nothing

I.2.23 Shop of Verecundus

Nothing.

Latrine under stairs to left (Eschbach p18: Fiorelli p37)

I.2.24,25,26 Workshop with shop

Nothing

Latrines in 24 & 26 (Eschebach p19: Fiorelli p37: Packer in 24 kitchen)

I.2.27.28.29 House and thermopolium of Polybius

Nothing

Latrine in 28 (Eschebach p19: Fiorelli p38)

I.2.8 Down Pipe

I.2.16 Down pipe

I.2.17 Down pipe

REGIO I. 3

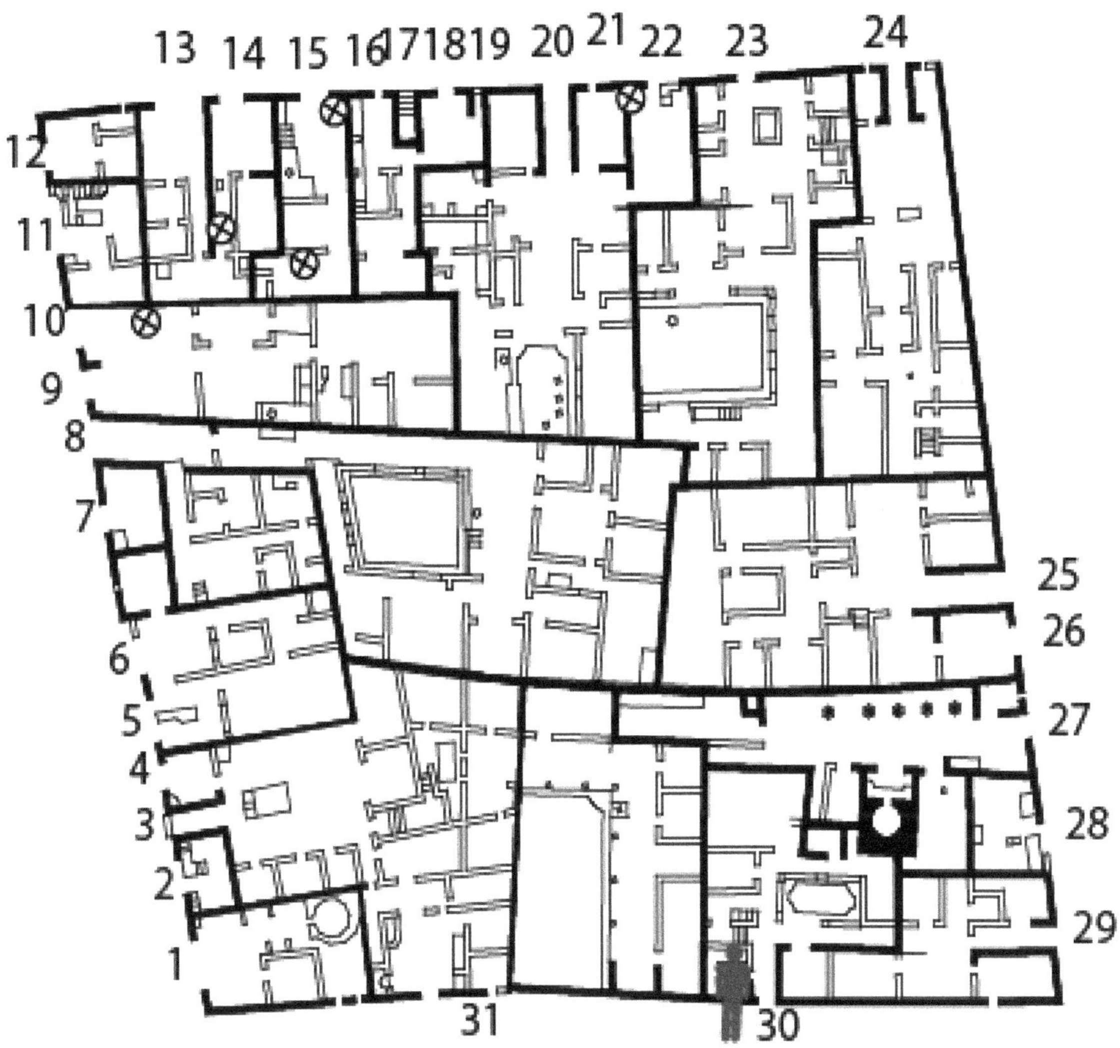

Twenty one properties. Eschebach and Fiorelli mention eight latrines. None can be found. There are five down pipes indicating some upper storeys at the north of the insula.

I. 3. 1 ***Pistrinum dulciarum***

I. 3. 2 ***Shop***

I. 3. 3, 4, 31 ***House and workplace of Epidius fortunatus***

Latrine in 31(Eschebach p20)

I. 3. 5, 6 ***Double shop***

I. 3. 7 ***Shop***

I. 3. 8 ***Entrance to house***

Nothing

Latrine under stairs (Eschebach p21): (Fiorelli p39)

Two possible situations

Nothing

I. 3. 9, 10 ***Shop and Workplace***

Latrine in kitchen in 10 (Eschebach p22)

I. 3. 11 ***Shop***

I. 3. 12 ***Shop***

I. 3. 13, 14 ***Two shops and workplace***

Nothing

Latrine (Fiorelli p39)

DOWN PIPE western wall of corridor, room f

I. 3. 15 ***Workshop***

Nothing

Latrine (Fiorelli p39)

I. 3. 16, 17, 18, 19 ***Laundry of Sestius Venustus***

I. 3. 20, 21, 22 ***Restaurant of Narziss***

DOWN PIPE

Latrine in 20 (Fiorelli p39)

I. 3. 23 ***House of Actius Anicetus***

I. 3. 24 ***House of Capella***

I. 3. 25, 26 ***Casa Des Guerrieri***

I 3. 27 ***Pistrinum***

Nothing

Latrine (Eschebach p25)

I. 3. 28 ***Thermopolium***

I. 3. 29 ***House of Innulus***

I. 3. 30 ***House of the Peristyle***

NICHE LATRINE room b

Latrine (Eschebach p25)

I.3.13,14 Down pipe, room c

I.3.11 Down pipe

I.3.15 North wall, room 1, Down pipe

I.3.20 room b. Down pipe

I.3.13,15 Pipe in internal wall, room b

I.3.21 North wall, room 1. Down pipe

I.3.30 Latrine

REGIO I. 4

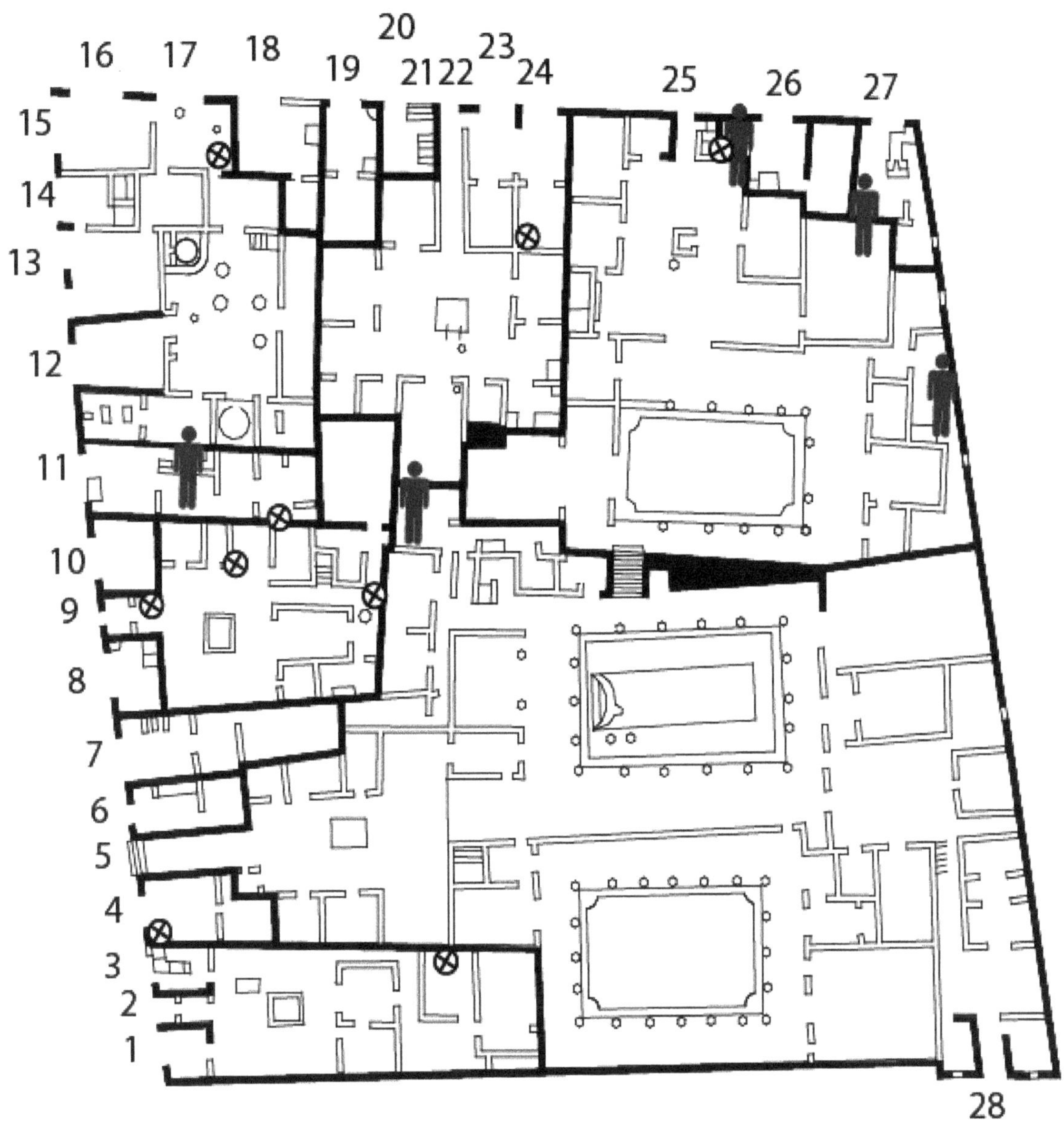

One enormous property, the *Casa del Citarista,* and thirteen much smaller ones make up this insula.

Two latrines in *Casa del Citarista.* Several down pipes suggesting upper storeys over the shops.

I. 4. 1, 2, 3 ***Restaurant***

DOWN PIPE North wall room f

I. 4. 4 ***Shop***

I. 4. 5, 6, 25, 28 ***House of L Popidius Secundus. Casa del Citarista***

LATRINE in I.4.25 room 64

LATRINE in I.5.28 room 42

I. 4. 7 ***Small fullonica of Passaratus and Maenianus***

Possible

I. 4. 8 ***Shop***

I. 4. 9 ***House of a worker in bronze***

Latrine (Fiorelli p43)

NARROW DOWN PIPE. May be for water to go into the cistern

DOWN PIPES in rooms a,e and n

I.4.10 Shop

Nothing

I. 4. 11 ***Caupona of Copiosus***

LATRINE in kitchen

Latrine (Eschebach p28): (Fiorelli p43)

DOWN PIPE in SE corner room d

I. 4. 12-17 ***Large establishment of D Junius Proculus***

I. 4. 18 ***Shop***

I. 4. 19 ***Shop***

I. 4. 20, 21 ***Shop***

I. 4. 22 ***Casa del Pressorio di Terracotta***

I. 4. 23, 24 ***Double shop***

I. 4. 26 ***Office***

DOWN PIPE

LATRINE

The latrine is to the left of the entrance. The down pipe is on the right. The floor has collapsed into the cess pit for the down pipe.

(Eschebach p30)

I. 4. **27** ***Thermopolium***

Latrine (Eschebach p31)

LATRINE in south western corner

I.4.4 SW corner. Down pipe

I.4.1,2,3 N wall room f. Down pipe

I.4.5,6,25,28 Room 42. Latrine

I.4.7 Down pipe

I.4.9 Fauces N wall. Down pipe

I.4.9 E wall room n. Down Pipe

I.4.11 Down pipe

I.4.11 Probable latrine

I.4.17 Down pipe

I.4.25 Room 64. Latrine

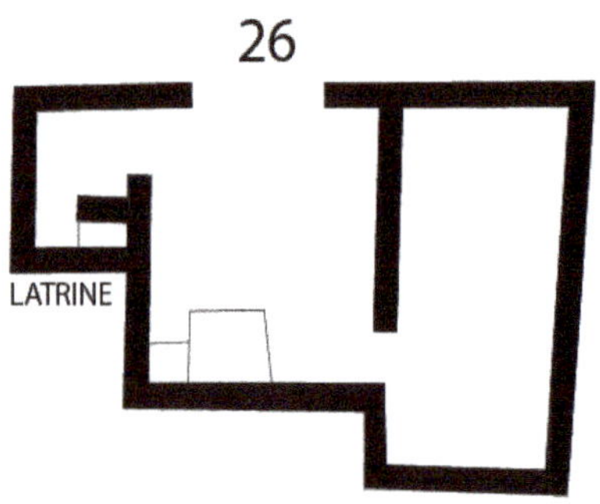

I.4.26 Latrine

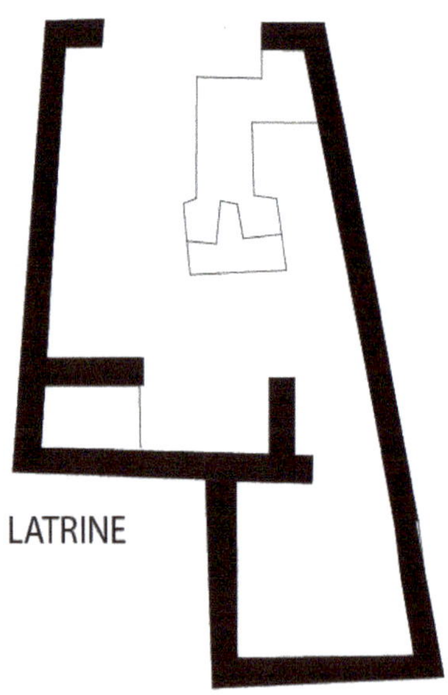

I.4.27

REGIO I. 5

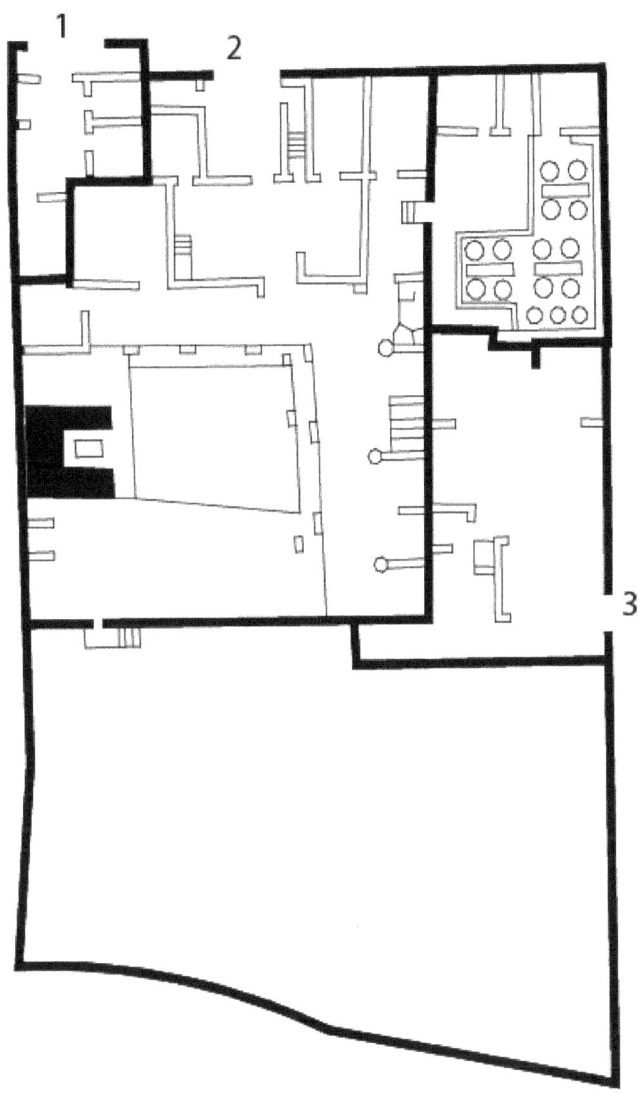

Only three properties in this insula and Eschebach does not describe any latrines

Virtually all of the insula is covered in undergrowth

I. 5. 1 ***House with Etruscan capitals***

NO ACCESS

I. 5. 2 ***House of M Vesonius Primus***

POSSIBLE LATRINE to the right of the entrance

I.5.3 Gardenhouse

I.5.2 Latrine in garden area h

REGIO I. 6

Eleven properties including four 'shops'.

Eschebach notes three latrines. The Fullonica of Stephanus is confusing. The 'shed in room l may or may not be the site of the latrine

I. 6. 1 ***Shop***

LATRINE in small room under the staircase

(Eschebach p32)

I. 6. 2. (4) 16 ***Complex House (House of the Cryptoporticus)***

Latrine in 2 over 1(Eschebach p33)

DOWNPIPE in room 29 on north wall

I.6. 3 ***Work area***

I.6. 4 Casa del Larario di Achille

I. 6. 5 ***Shop***

Nothing

I. 6. 7 ***Fullonica of Stephanus***

Latrine in a small 'room' adjacent to the kitchen (Eschebach p34)

The LATRINE is probably part of the kitchen, room m, and has a plethora of pipes in the wall

I. 6. 8. 9. 11 ***House of the Calavii***

I. 6. 10 ***Workshop***

(Modern custodi room)

I. 6. 12 ***Shop of Iunianus***

I. 6. 13. 14 ***House of Stallius Eros***

POSSIBLE by the staircase at the rear of the house

I. 6. 15 ***House of the Ceii***

LATRINE in kitchen to left of entrance, room I (Michel 1990.26)

Sloping tiles. Window

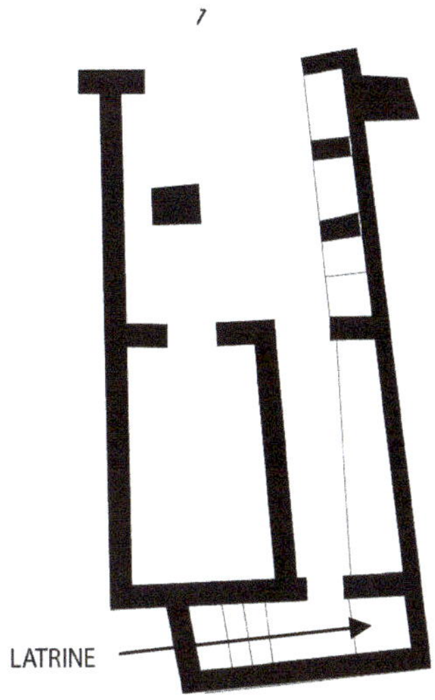

I.6.1

I.6.7 Latrine with several pipes

I.6.2 Room 29

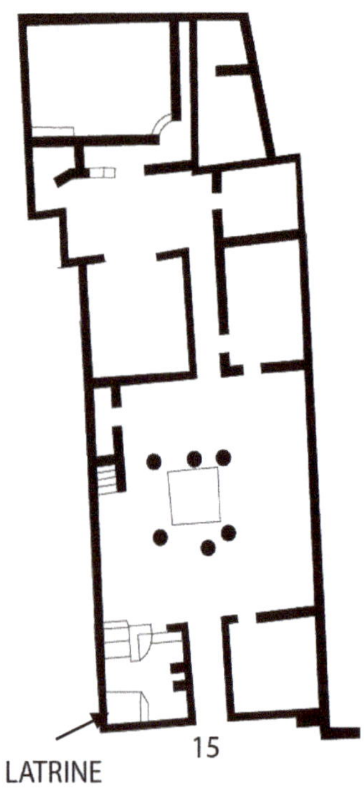

I.6.15

REGIO I. 7

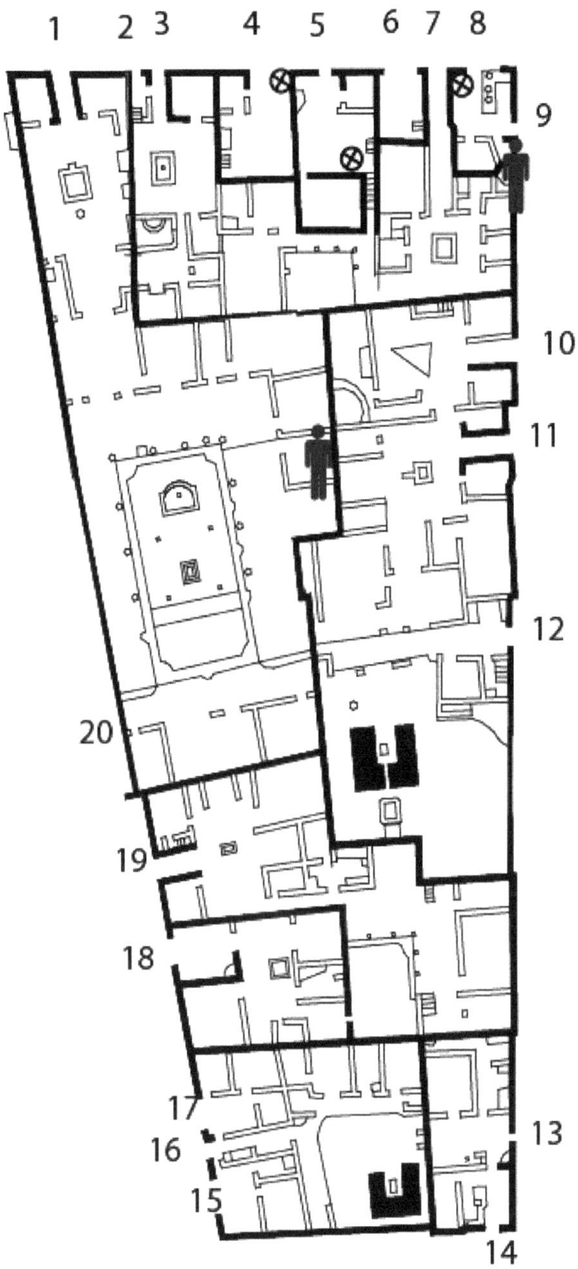

Eleven properties. Eschebach notes five latrines.

I.7.1,20 ***House of P Paquius Proculus***

TWO LATRINES

A complex group of latrines and down pipes at different levels, including a niche.in room 14

For the Latrine in room 36 (Ehrhardt 1998, 112)

I.7.2,3 ***House of M Fabius Amandio***

I.7.4 ***Shop***

DOWN PIPE in east wall, room 1

I.7.5 ***House of Philippus***

DOWN PIPE **in south wall of room c**

I.7.6 ***Shop of Primilla***

I.7.7 ***House of the priest Amandus***

LATRINE in north east corner to the south of I.7.9 room h

I.7. 8,9 ***Thermopolium***

Latrine in 8 (Eschebach p40)

DOWN PIPE at north western corner of property

I.7. 10-12,19 ***House of P Cornelius Tages***

Latrine in kitchen in 11 (Eschebach p40)

Latrine in kitchen in 12 (Eschebach p40)

LATRINE in Kitchen

WATER PIPE

I.7. 13,14 ***Caupona and brothel of Masculus***

The narrow room at the rear would be the likely situation but no convincing evidence

Latrine in kitchen in 13(Eschebach p41)

I.7. 15,16,17 ***Officina scriptorium***

POSSIBLE in 15

I.7. 18 ***Workplace of Niraemius***

Latrine in kitchen (Eschebach p41)

I.7. 19 ***House***

I.7.1 Latrine in lower room

I.7.1 Niche Latrine and Down Pipe

I.7.4 Down pipe north east corner

I.7.5 Down pipe

I.7.7 (2007)

I.7.8 Down pipe

I.7.12 Water pipe

I.7.11.12 Latrine

REGIO I. 8

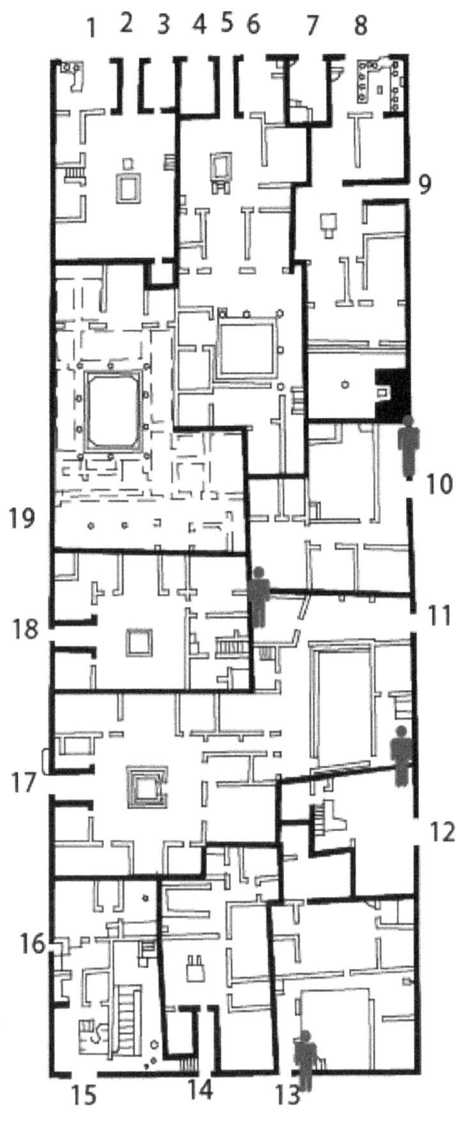

Twelve properties. Four latrines one of which is in the upper storey. Eschebach notes six latrines

I. 8. 1. 2. 3 *House of Stephanus*

I. 8. 4. 5. 6 *House of Chrysanthus*

I. 8. 7 *Shop*

POSSIBLE by staircase

I. 8. 8. 9 *Thermopolium of L Vetutius Placidus*

Latrine in 9 (Eschebach p44)

POSSIBLE to right of entrance

I. 8. 10 *Caupona of Pulcinella*

LATRINE under garden triclinium of 9 (Eschebach p44)

LATRINE in dark room to north of entrance.

I. 8. 12 *Workshop*

Possible to right of entrance

I. 8. 13 *Workshop of A Granius Romanus*

Latrine (Eschebach p45)

LATRINE to right of entrance

I. 8. 14 *House of M Epidius Primus*

Latrine (Eschebach p45)

NO ACCESS due to work in the street

I. 8. 15. 16 *Caupona of N Fufidius Successus*

Latrine in kitchen in 16 (Eschebach p46)

NO ACCESS due to work in the street

I. 8. 17. 11 *House of the four pens*

UPPER STOREY LATRINE

No Photograph

I. 8. 18 *House of Balbus*

POSSIBLE at rear of property

I. 8. 19 *Infectoria of Terentius*

Latrine (Eschebach p47)

NO ACCESS

I.8.10 Latrine

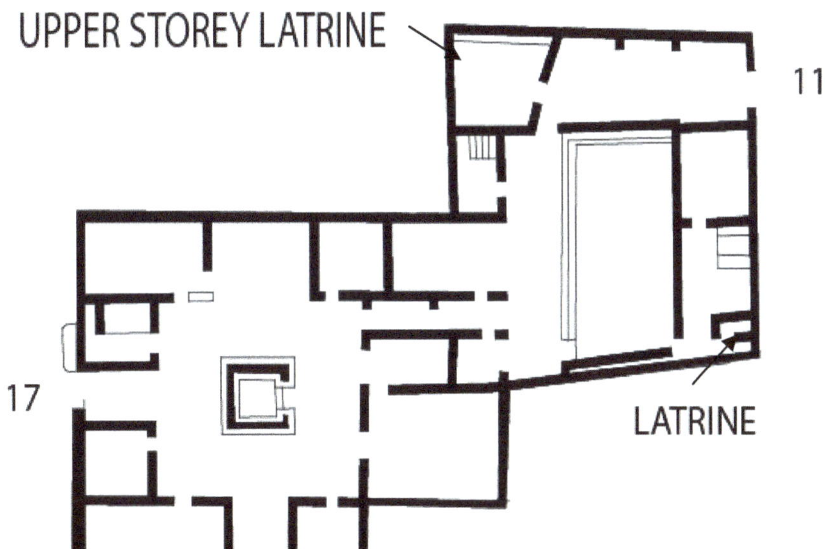

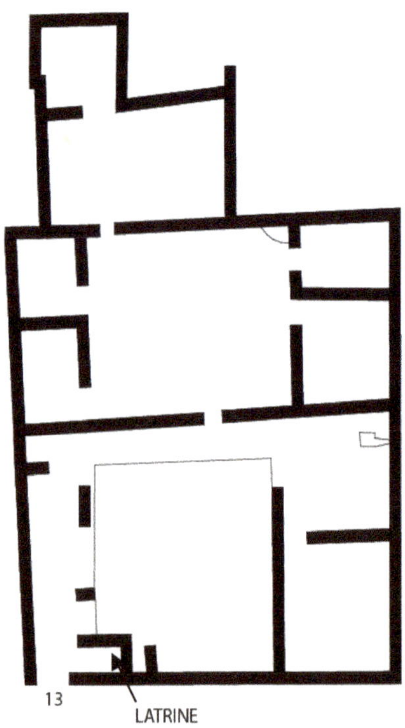

I.8.13

REGIO I.9

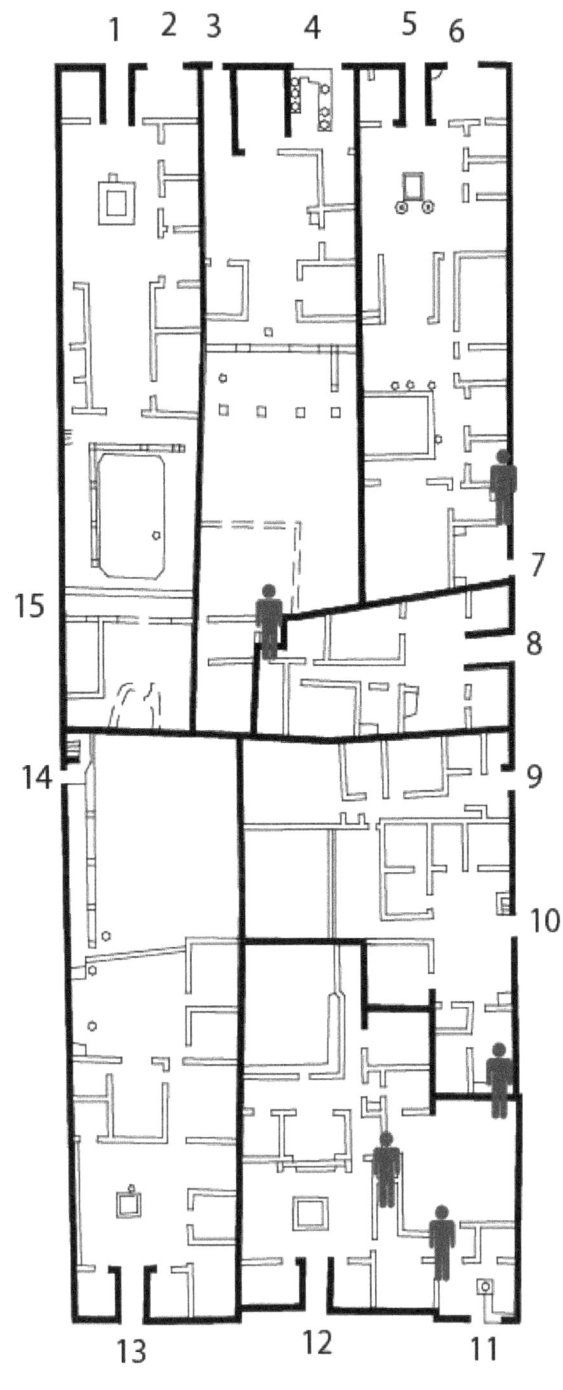

There are only eight properties in this insula. It has been possible to identify five latrines

I.9.1,2 ***House of the Beautiful Impluvium***

I. 9. 3, 4 ***House of Successus with thermopolium***

Latrine in 3 (Eschebach p48)

LATRINE in southern room

I. 9. 5, 6, 7 ***House of Euplia***

Latrine in 5 (Eschebach p49)

Nothing found in 5

Latrine in 7 (Eschebach p49)

LATRINE to north of entrance #7 room 15

The seat is modern and made of a plastic type material

I. 9. 8 ***Fish House of Romulus***

I. 9. 9, 10 ***Double house***

Latrine in 10 (Eschebach p50

LATRINE **in room 2**

I. 9. 11, 12 ***Caupona of Amarantus Pompeianus and brothel of Q Mestrius Maximus***

Two latrines

LATRINE and DOWN PIPE (A) in centre of property

LATRINE (B) for bar, to rear of doorway #11

I. 9. 13, 14 ***Casa di Cerere***

Latrine in 14 (Eschebach p51)

I.9. 15 ***Workshop***

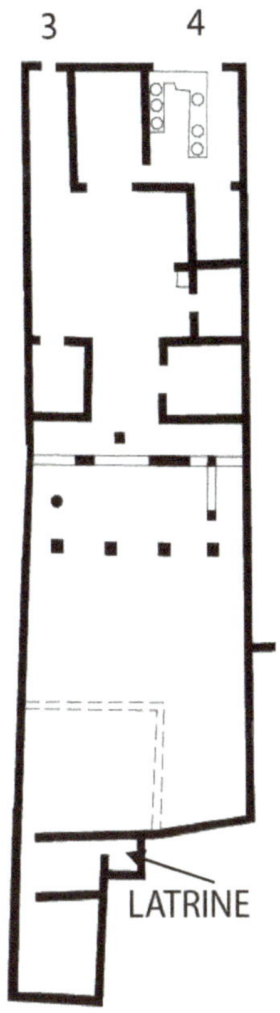

I.9.3,4 Latrine

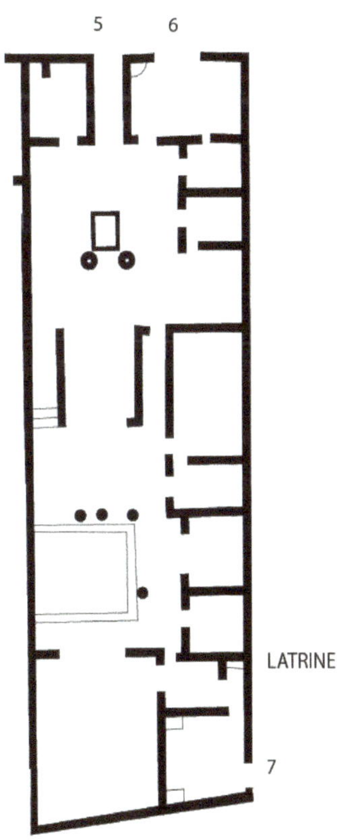

I.9.9,10 After cleaning

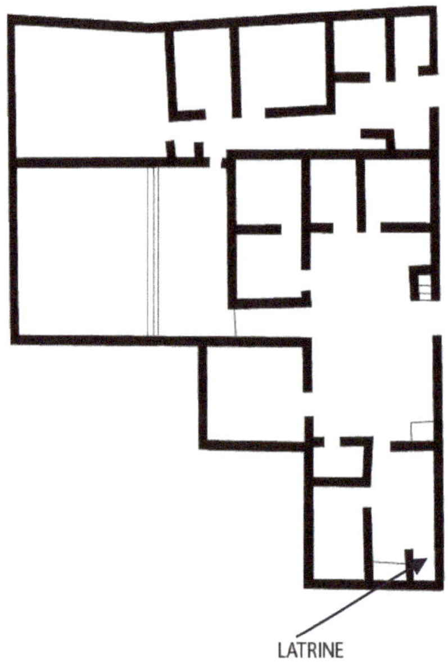

I.9.5-7 Latrine with modern seat

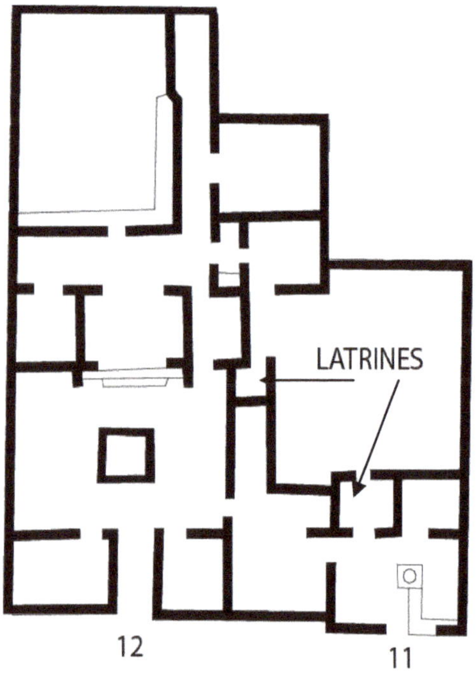

I.9.11,12 (A)

I.9.11,12 (B)

I.9.11,12A (2009)

REGIO I. 10

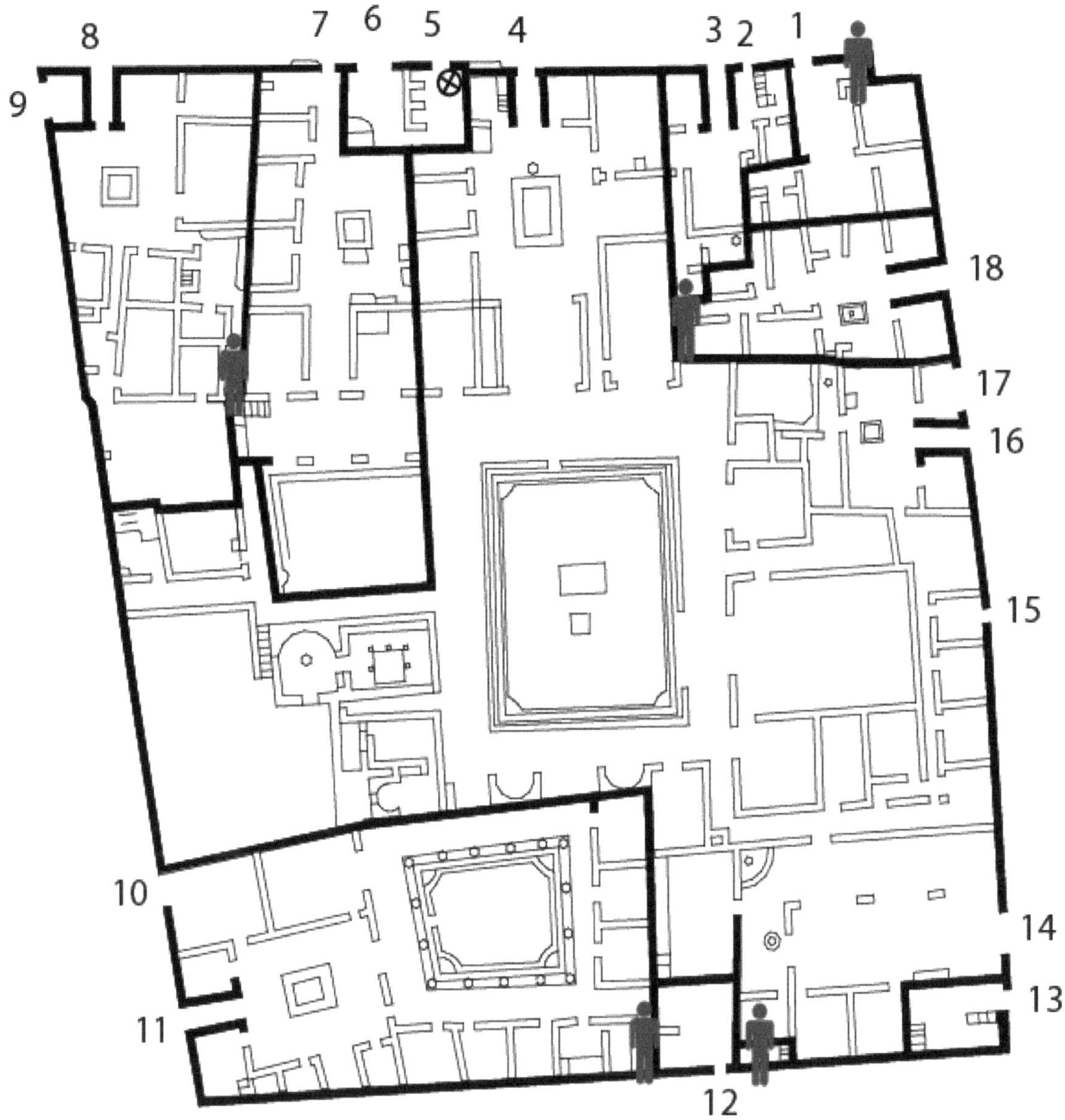

One very large property (house of Menander) with four modest houses and four shops make this an interesting insula. Five latrines have been identified and one down pipe.

I. 10. 1 ***Small workshop***

Latrine (Eschebach p52)

LATRINE in room to left of entrance

I. 10. 2. 3.*Caupona*

Latrine in kitchen in 3 (Eschebach p52)

LATRINE/DOWN PIPE

I. 10. 4. 14-17 ***House of Menander***

Latrine in peristyles area in 4 (Eschebach p53)

Latrine in 16 (Eschebach p53)

LATRINE in room 31

I. 10. 5 ***Brothel***

DOWN PIPE

I. 10. 6 ***Workshop***

Pit for upper storey Latrine (Eschebach p54)

I. 10. 7 ***House of M Volusius Iuvencus***

Latrine to right of entrance (Eschebach p54)

Possible

I.10. 8 ***House of Minucius***

Latrine in kitchen (Eschebach p55)

LATRINE in room 14

I. 10. 9 ***Workshop***

I. 10. 10, 11 ***House of Ti. Claudius Eulogos***

Latrine in 11 (Eschebach p56)

LATRINE in room 14

I. 10. 12 ***Workshop***

Nothing

I. 10. 13 ***Caupona***

Nothing

I. 10. 18 ***House of Aufidius Primus***

Latrine (Eschebach p57)

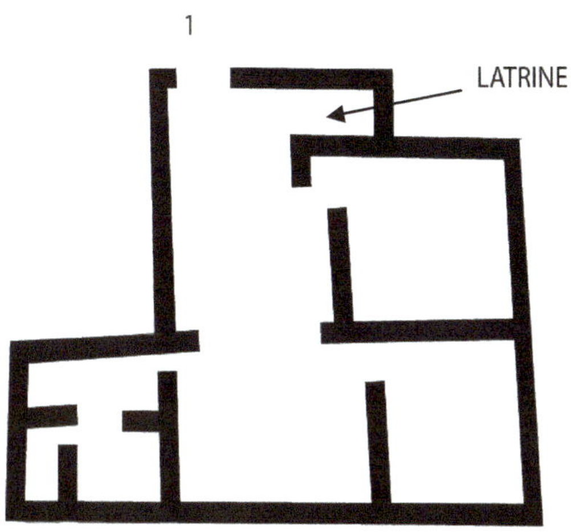

I.10.1

I.10.2,3 Down pipe

I.10.4 House of Menander: Room 31

I.10.6 Upper storey latrine with downpipe in masonry

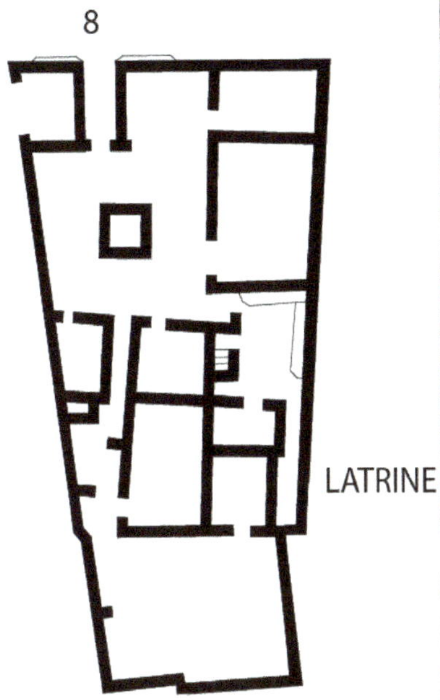

I.10.8 Latrine and Down pipe: Room 14

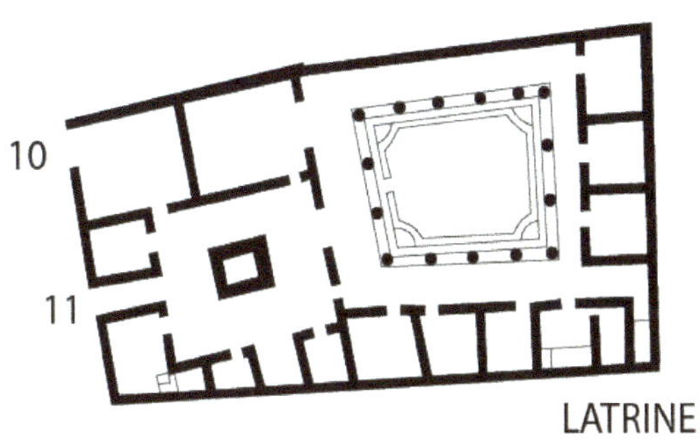

I.10.10,11: Room 14

REGIO I. 11

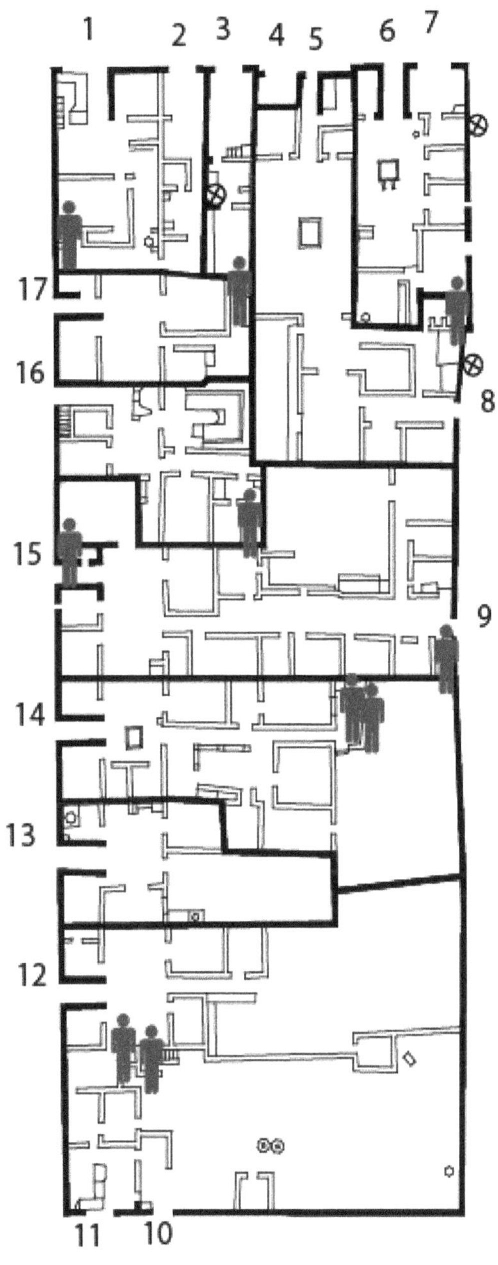

Eleven properties in this insula including one shop

Nine latrines have been identified and also two down pipes in the outer eastern wall of the insula

I.11. 1, 2 *Caupona Stabilionis*

LATRINE in I.11.1, room 7

This LATRINE is set against the western wall

Latrine in I.11.2 (Eschebach p58)

I.11. 3 *Shop or workshop*

Latrine in kitchen (Eschebach p58)

I.11. 4 *Small shop*

I.11. 5, 8 *House of Lollius Synhodus*

Latrine in 8 (Eschebach p58)

Niche LATRINE against eastern wall

DOWNPIPE in outer wall

I.11. 6, 7 *House of Venus in the bikini*

Latrine in 6 (Eschebach p59)

Nothing found in the house but a DOWN PIPE in the outside wall

I.11. 10, 11, 12 *Caupona of Euxinus and Iustus*

Latrine in 10 (Eschebach p59)

This LATRINE is accessed from the garden and has a small water tank

Latrine in 11(Eschebach p60)

This LATRINE is accessed from the house and is side by side with the garden latrine.

I.11. 13 *House*

I.11. 14 *House of Cherem*

LATRINE at rear of property, room o.

May be two-seater

I.11. 15, 9 *House of the superior piano*

Latrine in 9 and 15 (Eschebach p61)

LATRINE in room 16

LATRINE in room 1

I.11. 16 *Inn with restaurant*

Latrine (Eschebach p61)

Tiled floor. Probable LATRINE in room 10

I.11. 17 *Imperial House*

Latrine (Eschebach p61)

LATRINE in north-east corner, room 8

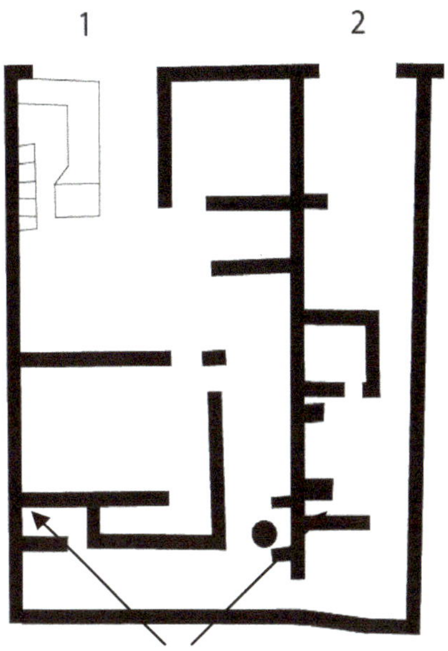

I. 11. 1

I. 11. 2

I. 11. 3

I. 11. 7 Down pipe in outside wall

I.11.8 Niche latrine

I. 11. 8 Down Pipe in outer wall

This down pipe is further south than the latrine in the property and may not drain into the same cesspit

I. 11. 10

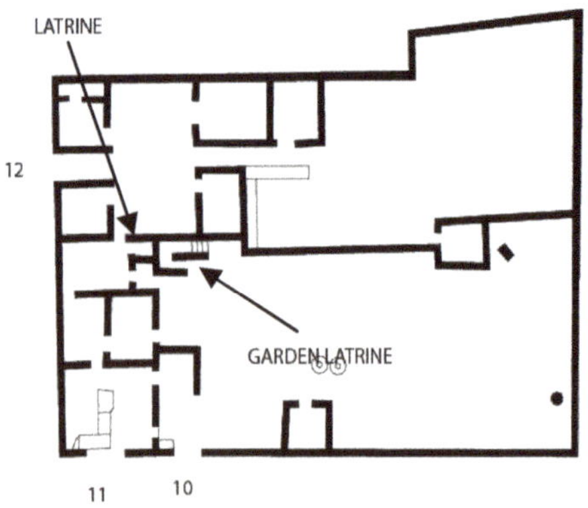

I. 11. 11

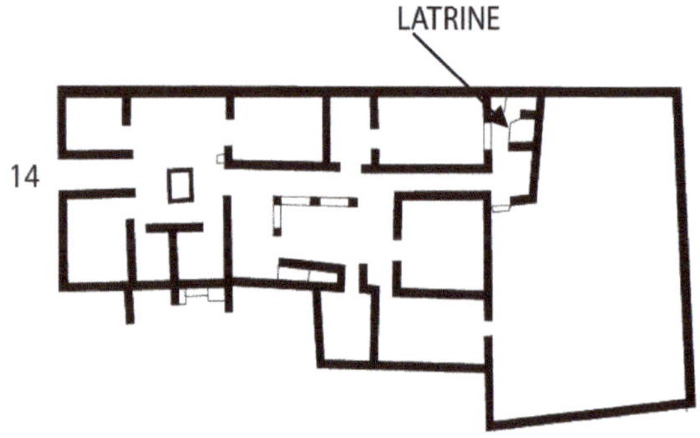

I. 11. 14 Two-seater latrine

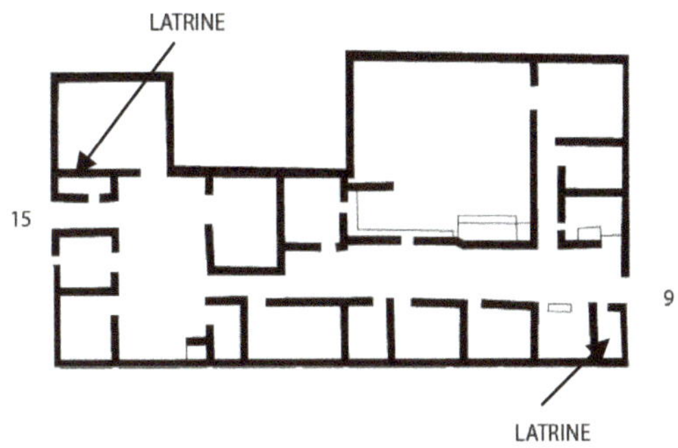

I. 11. 15

I.11.15

I. 11. 16

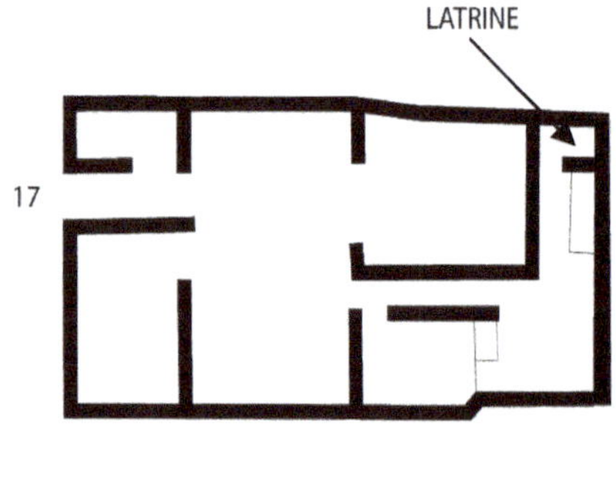

I. 11. 17

REGIO I. 12

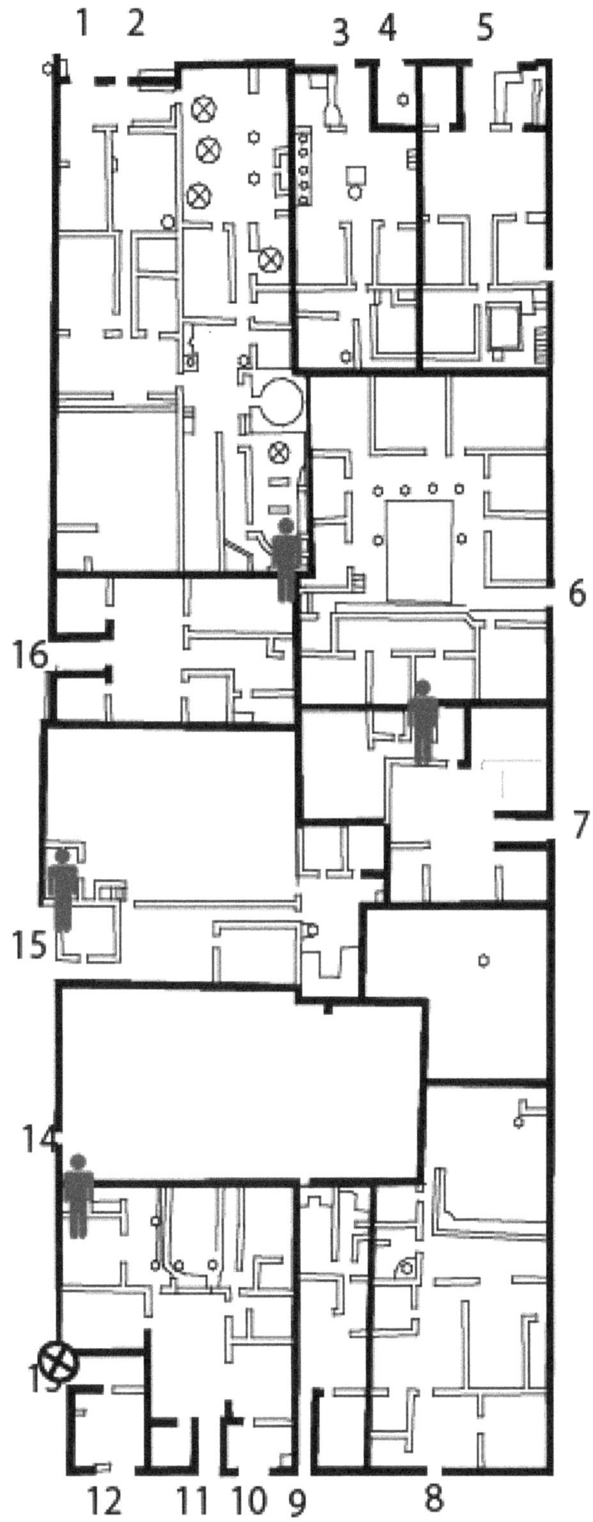

Twelve Properties including two small shops. Three latrines and one other possible.

One down pipe

I.12. 1. 2 *Pistrinum of Sotericus*

Latrine in 2 (Eschebach p62)

LATRINE in 2

I.12. 3 *Restaurant of Sotericus*

Nothing

I.12. 4 *Shop*

Nothing

I.12. 5 *Caupona of Lutatius*

POSSIBLE. Tiled floor in narrow room in south west corner of property

I.12. 6 *Townhouse*

Nothing

I.12. 7 *House*

Latrine (Eschebach p64)

LATRINE in room 7

I.12. 8 *House of garum*

Latrine (Eschebach p64)

I.12. 9, 14 *Small house with large garden*

Latrine in kitchen in 9 ((Eschebach p65)

Nothing

I.12. 10, 11 *House with paradise picture*

LATRINE in kitchen area, room 11

I.12. 12, 13 *Bottega*

DOWN PIPE to North of Doorway 13

I.12. 15 *House of Medusa*

LATRINE in room 8

I.12. 16 *House*

Latrine in kitchen (Eschebach p66)

Nothing

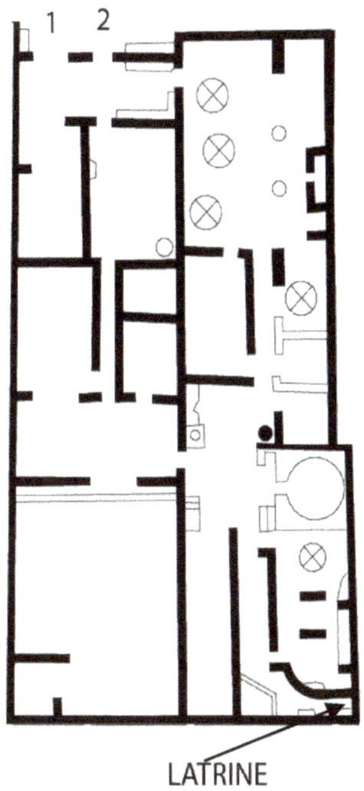

I.12.2

I.12.5

I.12.7

I.12.11

I.12.13 Downpipe

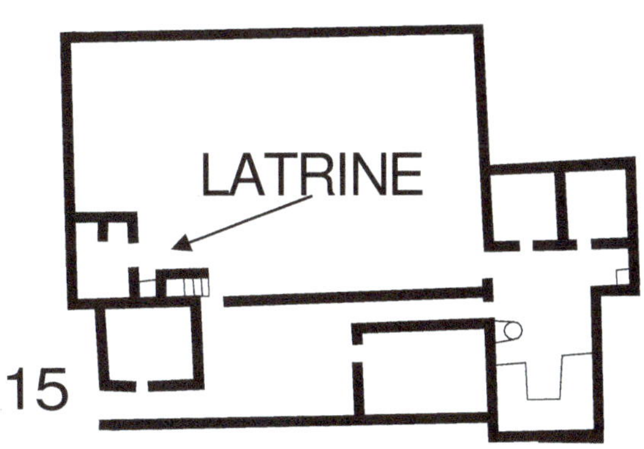

I.12.15 Latrine

REGIO I. 13

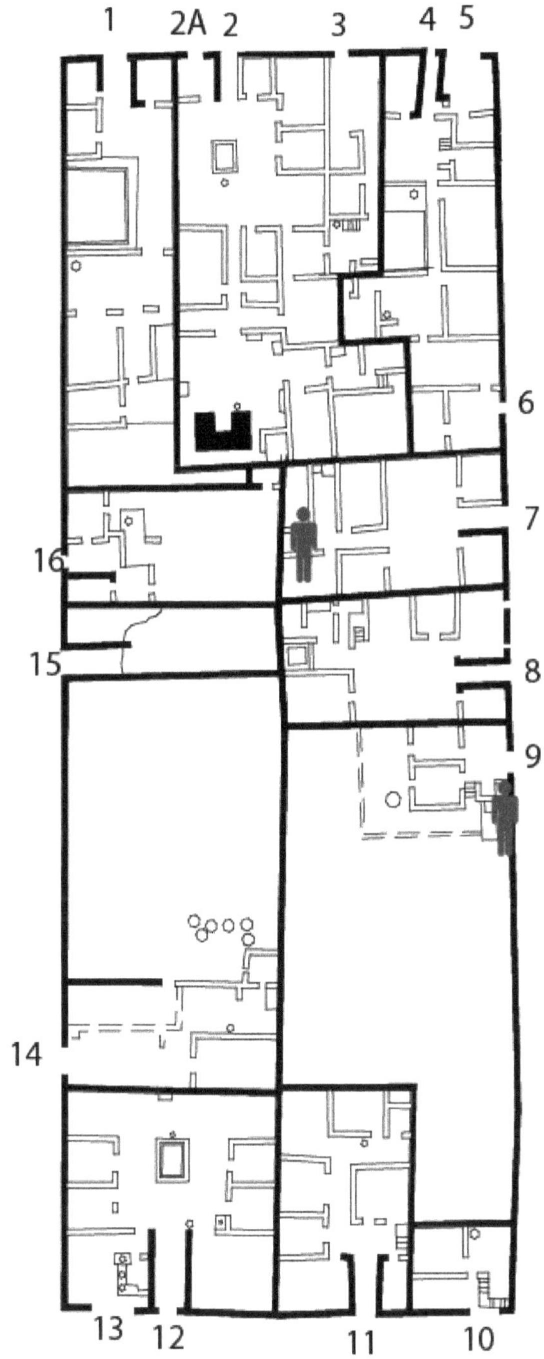

Another insula with twelve properties. Two latrines; no down pipes.

I. 13. 1 ***House of Crassius Crescens***

Latrine (Eschebach p67)

Nothing

I. 13. **2, 2a** ***House of Sutoria primigenia***

Nothing

I. 13. 3 ***Workshop of Lesbianus***

Latrine (Eschebach p68)

Nothing

I. 13. 4, 5, 6 ***Residence of Taedia Secunda***

Latrine in 4 (Eschebach p68)

Nothing

I. 13. 7 ***Restaurant***

LATRINE in rear of property

I. 13. 8 ***Private house***

Latrine ((Eschebach p68)

NO ACCESS

I. 13. 9 ***House of Venus Marina***

LATRINE in room 2

I. 13. 10 ***Small shop***

Nothing

I. 13. 11 ***Town house***

NO ACCESS

I. 13. 12-14 ***House of the Lararium of Isiac***

NO ACCESS

I. 13. 15 ***?Shop***

Possible

I. 13. 16 ***Caupona***

Latrine (Eschebach p71)

NO ACCESS

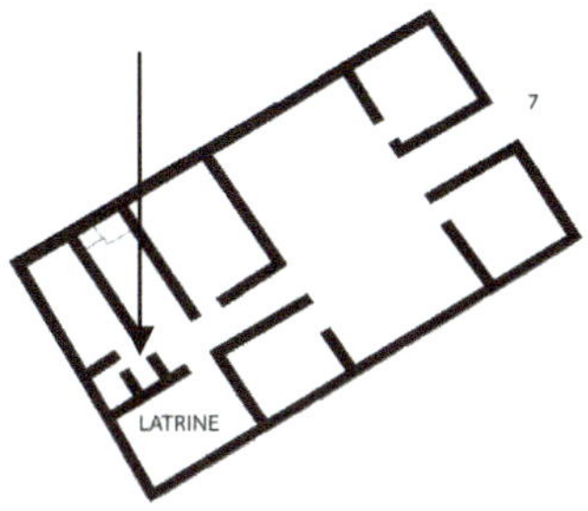

I. 13. 7

I. 13. 9 Room 2

REGIO I. 14

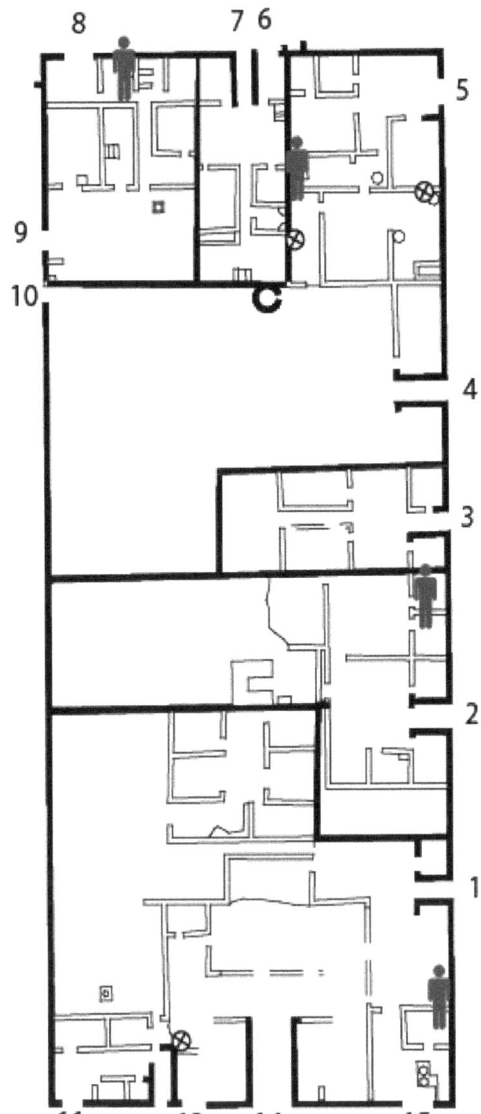

Only six properties in this insula. There are four latrines although Eschebach notes five. All the down pipes appear to be supplying water. The plan is not accurate for numbers 1 and 15, causing some confusion.

I. 14. 1, 11-15 *Inn of Astylus*

Latrine in 12 (Eschebach p71)

Latrine in 15 (Eschebach p74)

I. 14. 2 *Stabulum*

LATRINE in room 8, modified by insertion of amphora. This might be in order to conserve water.

I. 14. 3. *Small private house*

Latrine. (Eschebach p72)

I. 14. 4, 5, 10 *? Inn*

Possible LATRINE in 4.

Latrine in 5. (Eschebach p72)

LATRINE in 5.

I. 14. 6, 7 *House of the lararium of Sarno*

Latrine in kitchen in 7 (Eschebach p73)

Nothing.

I. 14. 8, 9 *Workshop or caupona*

LATRINE in room 2 to left of entrance.

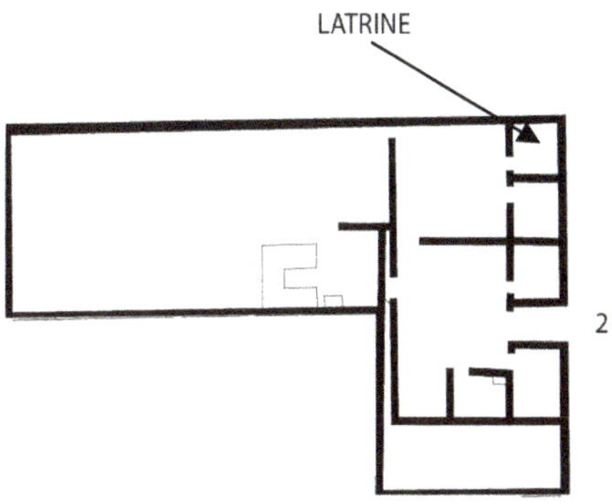

I. 14. 2

I. 14. 4

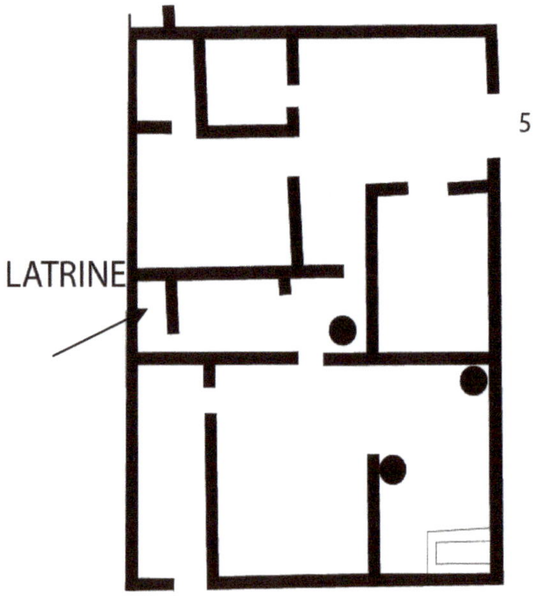

I. 14. 5

I. 14. 8, 9

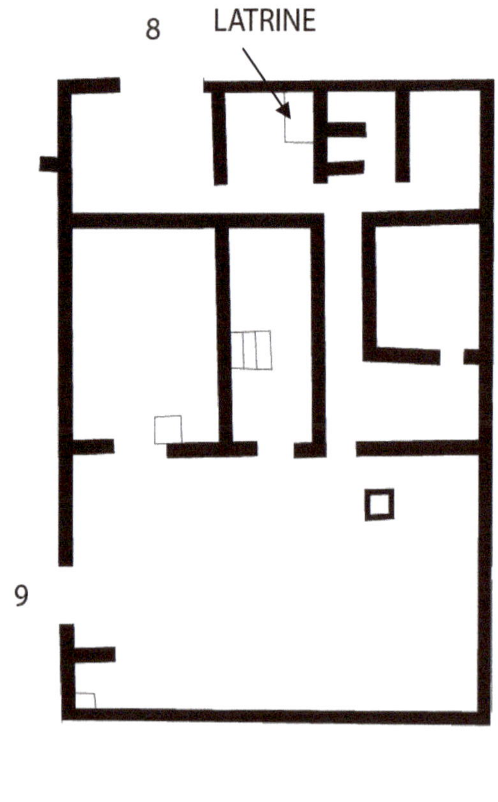

REGIO I. 15

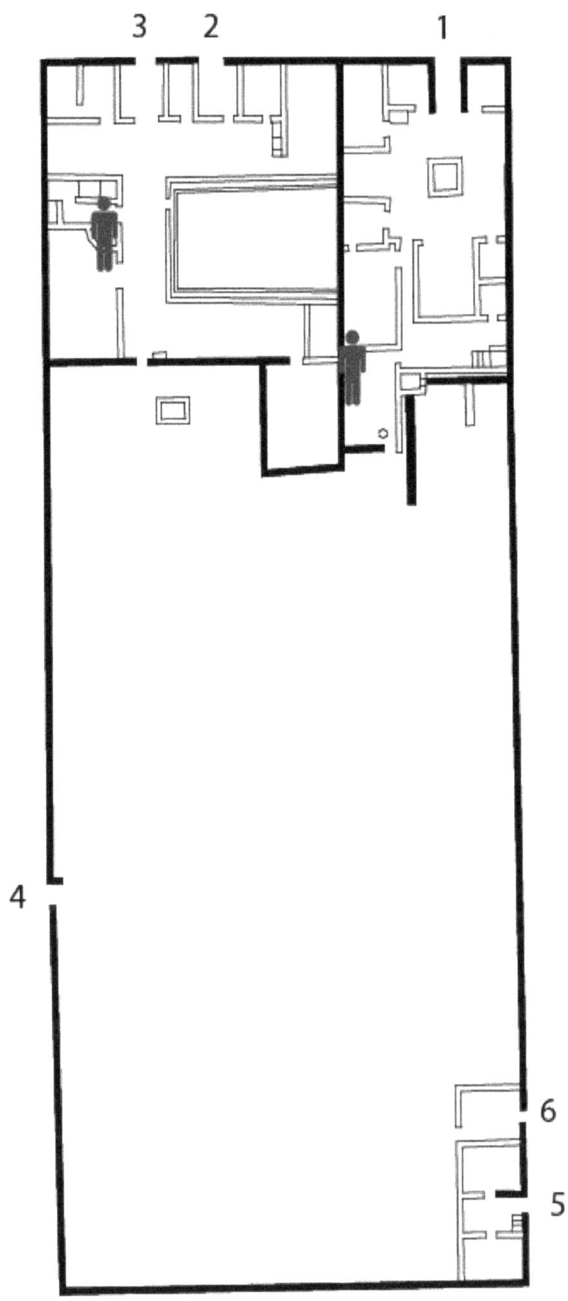

An insula with only two properties, each with a latrine

I. 15. 1 *Atrium house*

Latrine (Eschebach p74)

LATRINE at rear of house

I. 15. 2, 3, 4, 6 *House of the ship Europa*

Latrine in 3 (Eschebach p74)

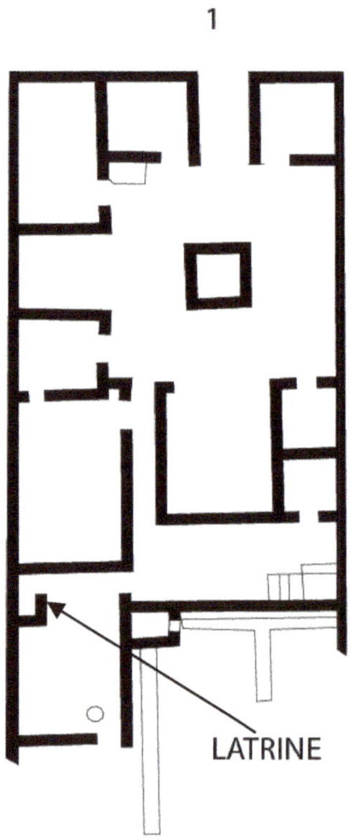

I. 15. 1

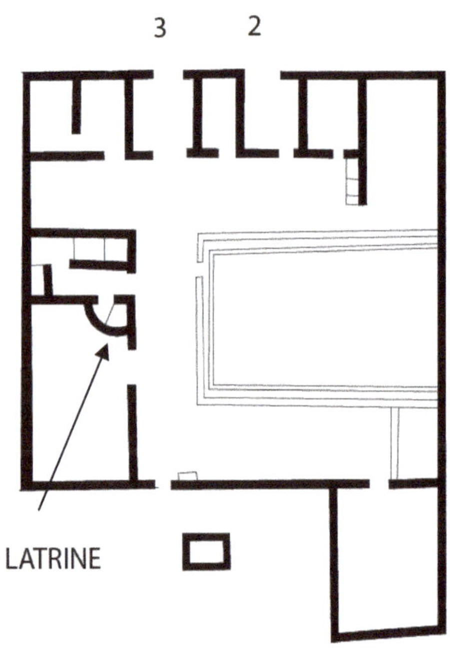

LATRINE in small quarter-circle shaped room 9

I. 15. 2, 3, 4, 6

REGIO I. 17

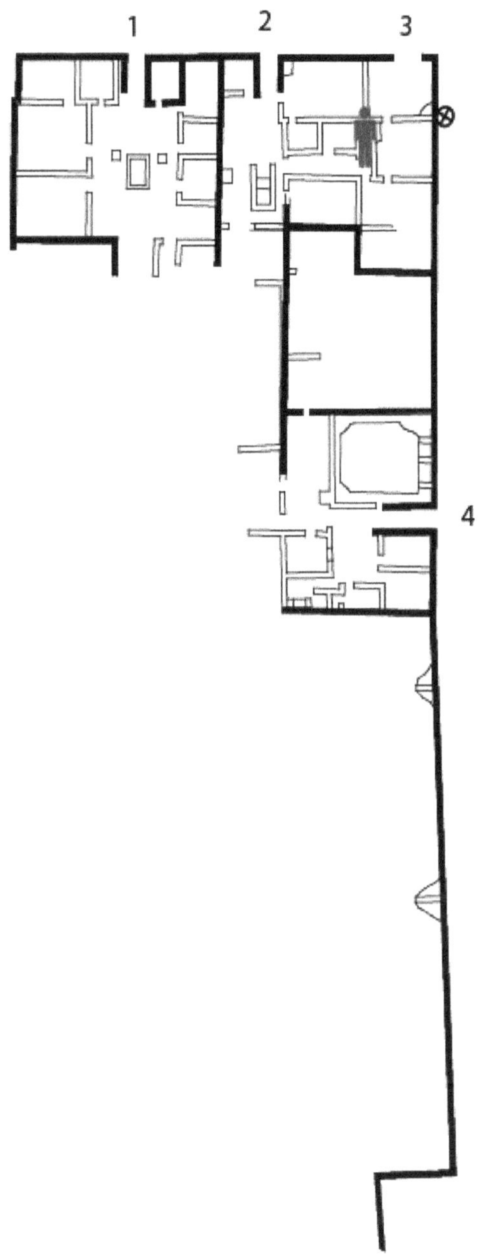

Only three properties in this partially excavated insula

I. 17. 1 ***Atriumhouse***

Nothing

I. 17. 2. 3 ***Inn***

LATRINE in 2 in centre of property

?Waste disposal chute in south east corner of room with door #3

DOWN PIPE in outer wall

I. 17. 4 ***House of the arches***

Latrine in kitchen (Eschebach p78)

NO ACCESS

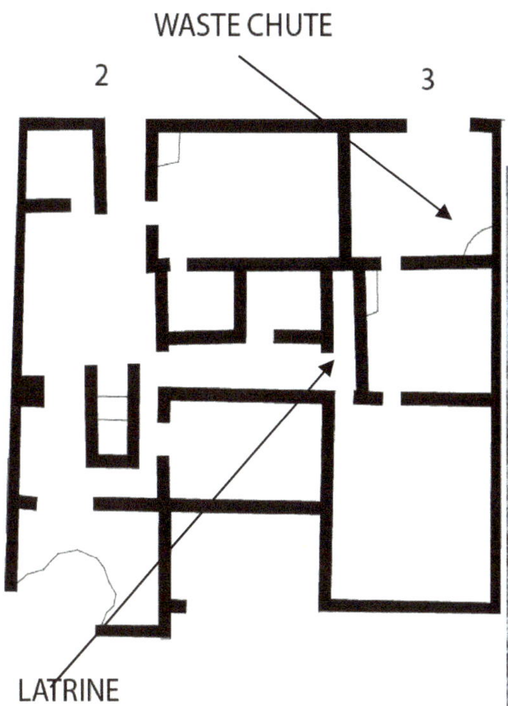

I. 17. 3 ?Waste disposal chute

I.17.2,3 Latrine

I.17.3 Down pipe in outer wall

REGIO I. 21

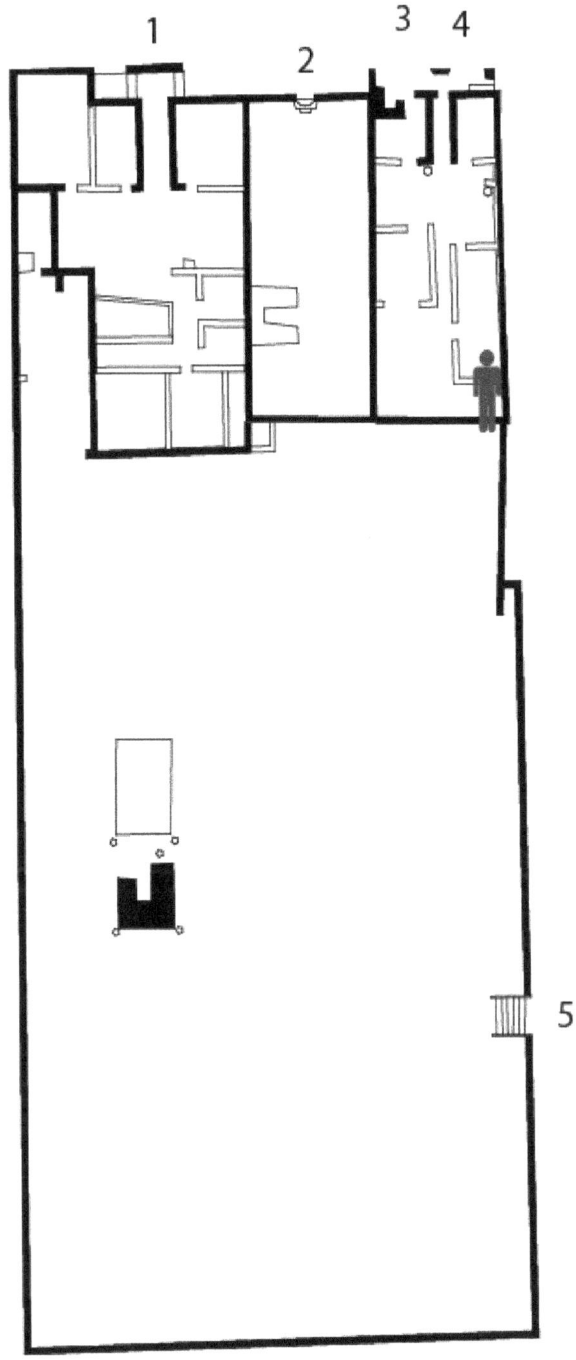

The insula with the House of the Fugitives. One latrine but an interesting feature probably associated with rubbish disposal

I.21. 1 ***House of the Fugitives***

A very interesting feature in the north of the garden of the House of the fugitives. Almost certainly used for food preparation/disposal

I.21. 2 ***Space with triclinium***

Nothing

I.21. 3-5 ***Workshop and Garden***

LATRINE in south-east room

Waste disposal feature in garden of the House of the fugitives

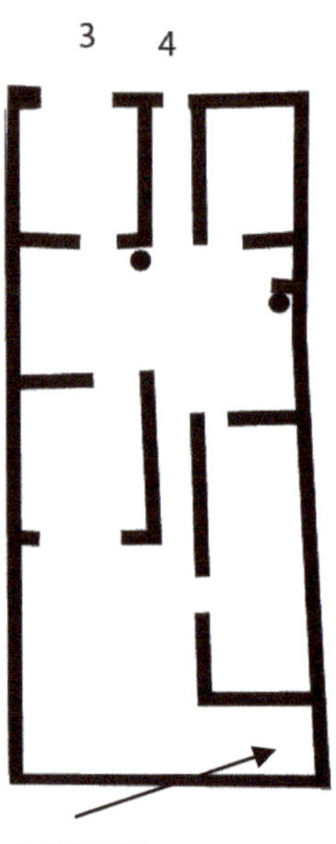

I. 21. 3, 4, 5

REGIO II

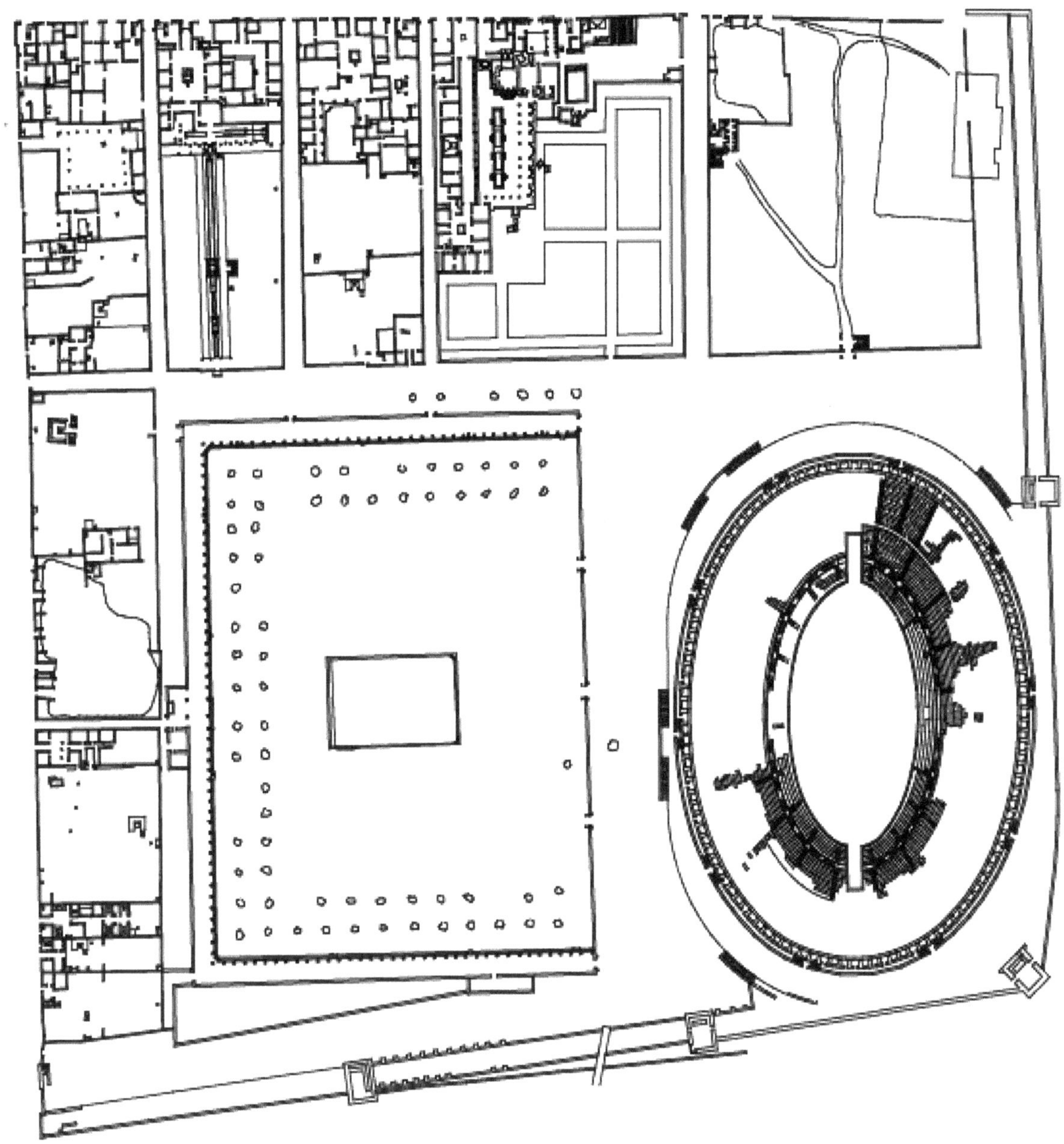

An interesting Region containing the amphitheatre and Grand Palaestra. The largest insula is mainly the Praedia Julia Felix.

REGIO II.1

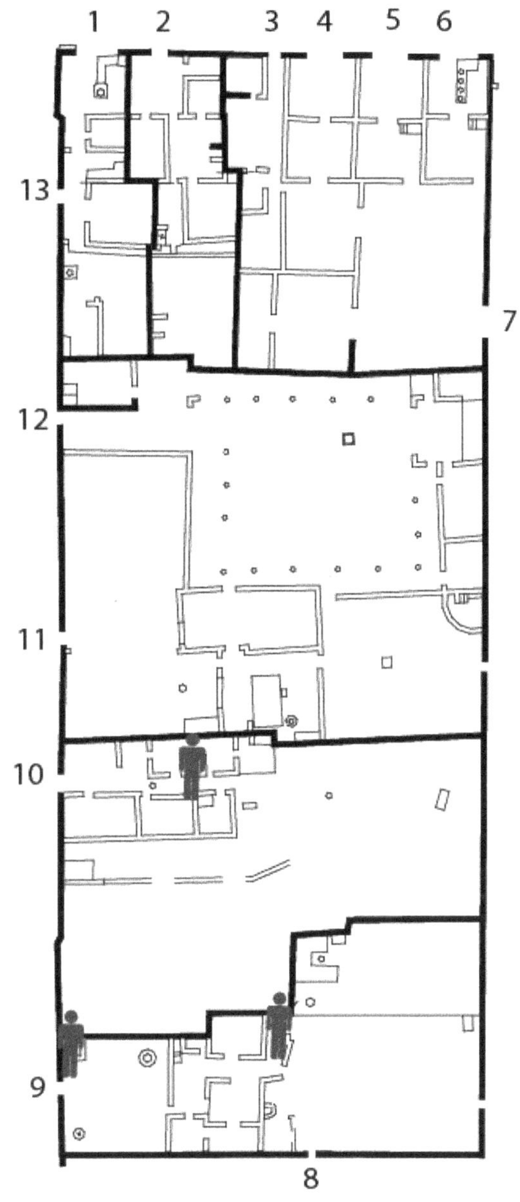

Only six properties with extensive garden areas.

II.I.1,13 ***Thermopolium and caupona of Hermes***

Nothing

II.I.2 ***House of the Aemilii***

Nothing

II.I.3-7 ***Workhouse with thermopolium***

Nothing

II.I.7,8,9 ***House of Felix and Sabinus***

Latrine in 9 (Eschebach p86)

LATRINE by entrance #9 room 1

LATRINE accessible from garden and triclinium #8 room 1

II.I.10 ***Casa Imperiale***

Latrine (Eschebach p87)

LATRINE off corridor, room 6

II.I.11,12,7a ***Cult Garden of Jupiter Sabazius***

Nothing

II.1.9,8,7

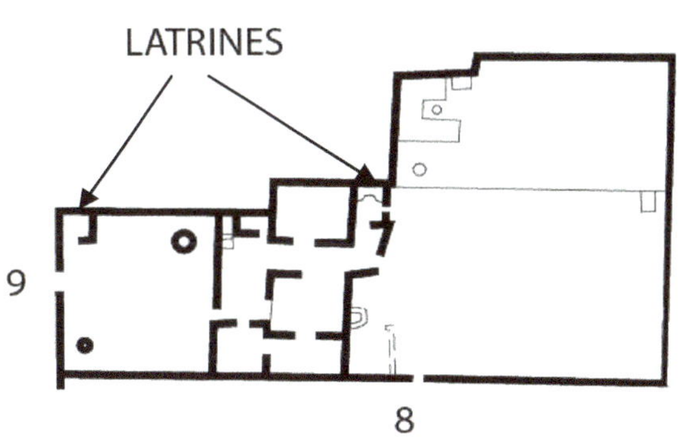

II.1.9,8,7

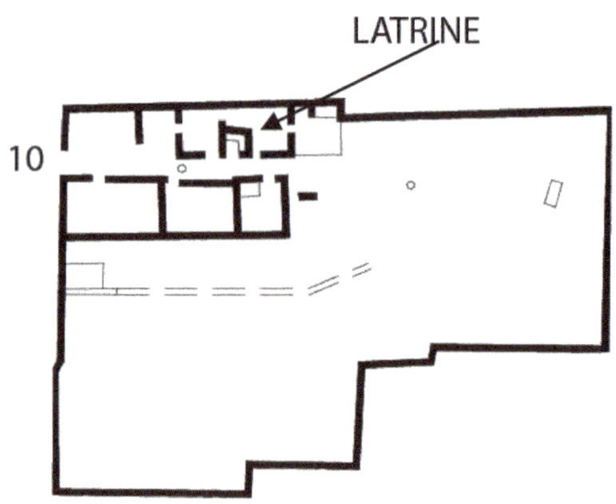

II.1.10

REGIO II.2

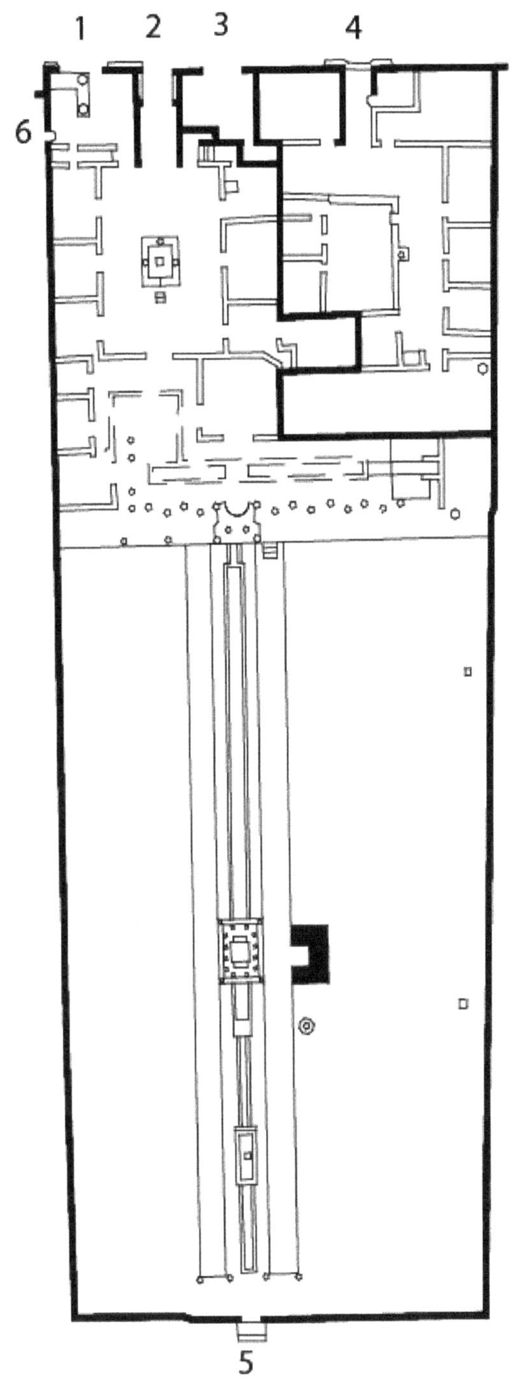

Insula of two properties. One possible latrine

II.2.1,2,3,5,6 ***House of Loreius Tiburtinus***

Latrine in restaurant room (Eschebach p889)

Nothing

Possible in **3 *Popina***

II.2,4 ***House of Messius Ampliatus***

?Latrine under Staircase (Eschebach p89)

Nothing

II.2.3 Possible latrine

REGIO II.3

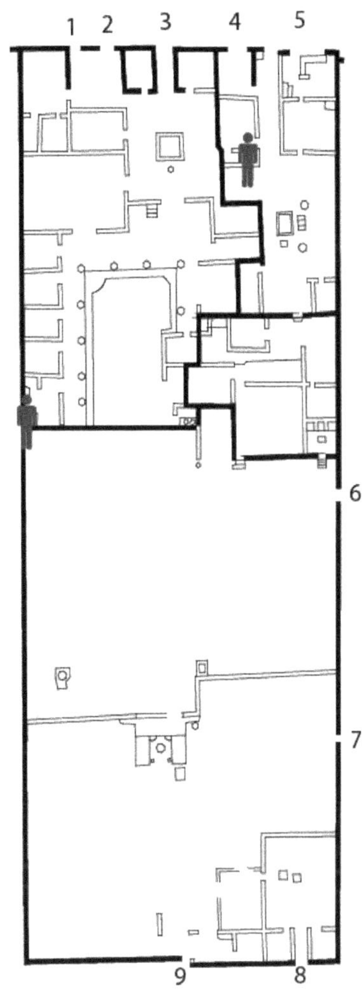

The House of Venus in the Shell is well presented to the public, except for the kitchen area which includes the latrine

II. 3. 1, 2, 3 ***House of Venus in the Shell***

LATRINE in kitchen

II. 3. 4, 5 ***Workhouse with large garden***

Probable latrine

II. 3. 7, 8, 9 ***Garden restaurant of Nicanor***

Nothing

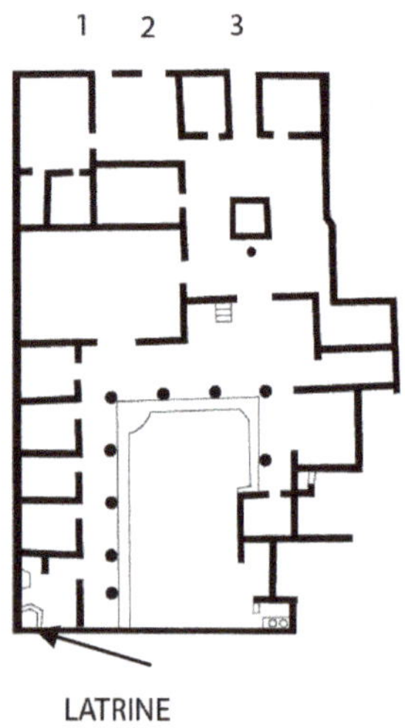

II.3.1,2,3

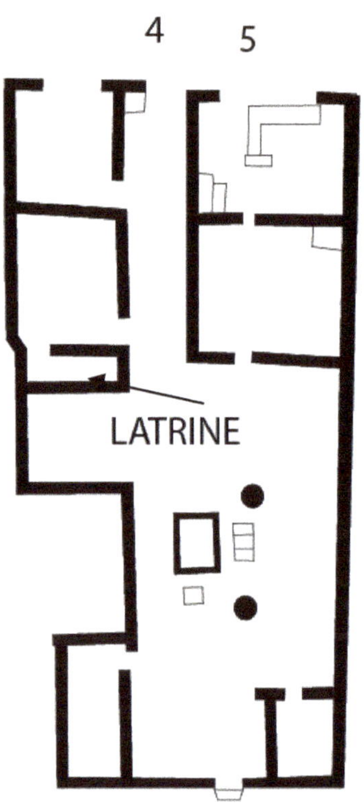

II. 3. 4, 5, 6

REGIO II.4

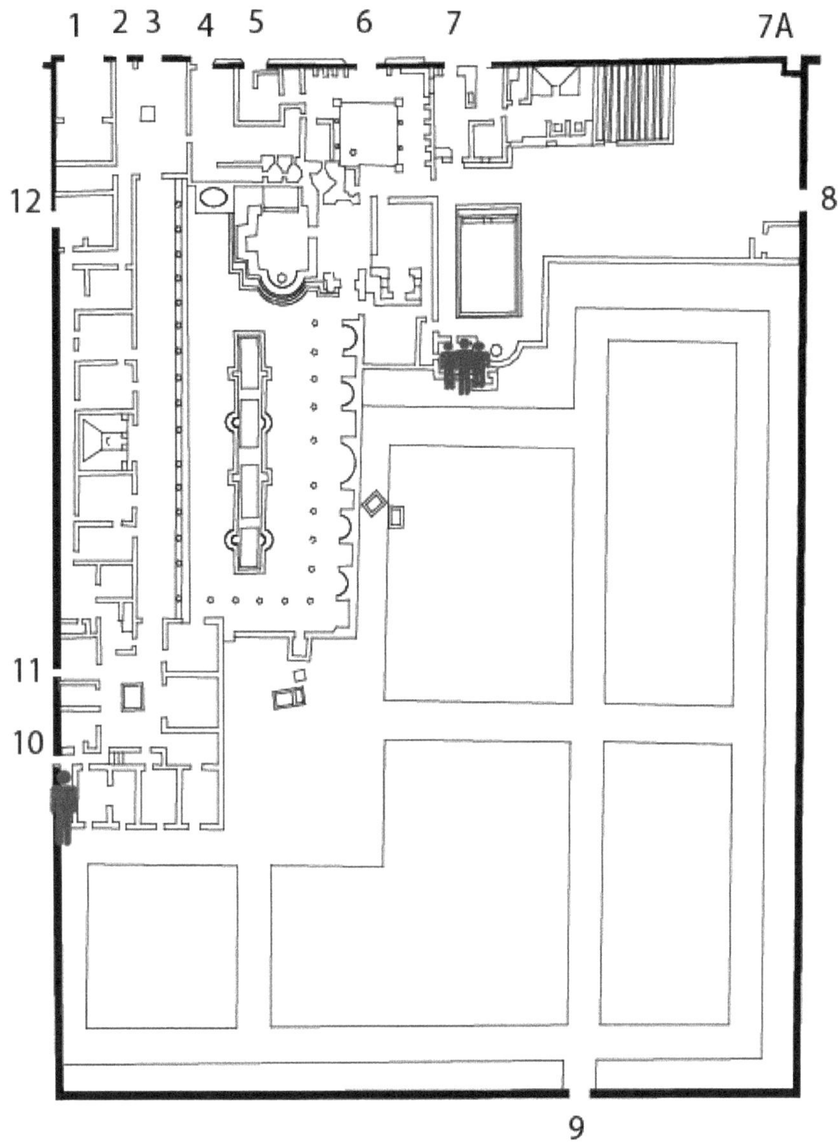

This insula is occupied by the Praedia of Julia Felix.

Two latrines are visible. There is a multi seat latrine which is part of the bath complex (room 7) (Eschebach p93) and a single latrine in the kitchen area by #10 (room 56)

(Eschebach p93). The water supply for this latrine is behind the rear wall of the latrine and may have been accessible from the kitchen.

Multi seat latrine: Praedia Julia Felix

II.4.10 Latrine in kitchen

REGIO II.7

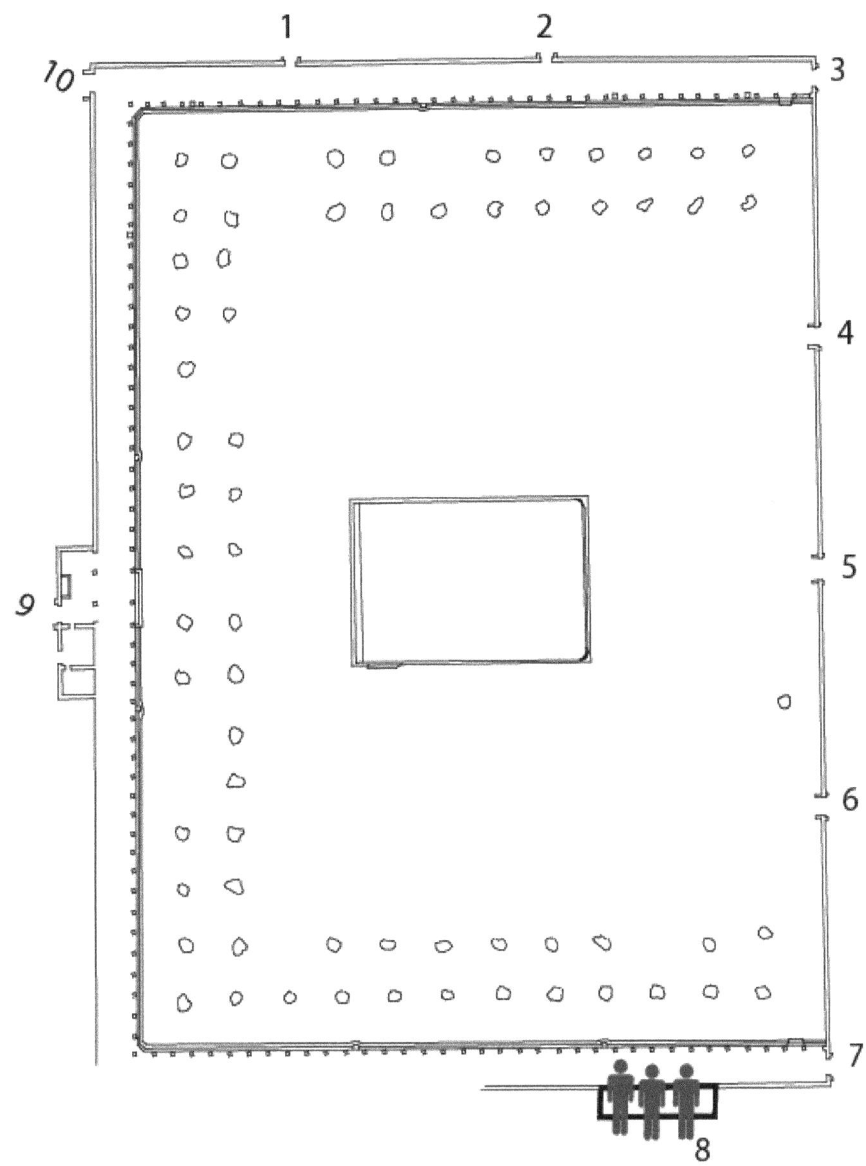

II.7.8

Multi seat latrine. (Eschebach p95) This building is used for storage. The windows are barred and even were access possible no descriptive photograph would result.

II.7.8 External view of latrine building from southeast

REGIO II.8

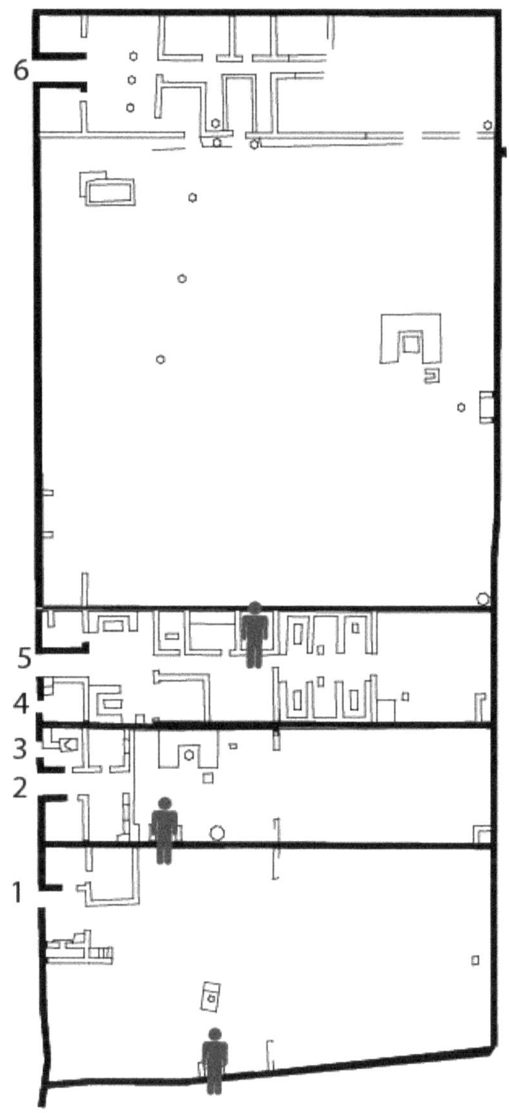

Four properties, probably associated with dining and entertainment. Three with latrines, two of which are in garden areas.

II.8.1 ***Food restaurant***

LATRINE by South wall, room 4

(Eschebach p96)

II.8.2,3 ***Thermopolium and restaurant***

LATRINE in small room within garden area, room 5

Latrine in 2 (Eschebach p96)

II.8.4 ***Shop***

Nothing

II.8.5 ***Restaurant with shop***

LATRINE on Left, room 8

Latrine in 5 (Eschebach p96)

II.8.6 ***Garden of Hercules***

Nothing

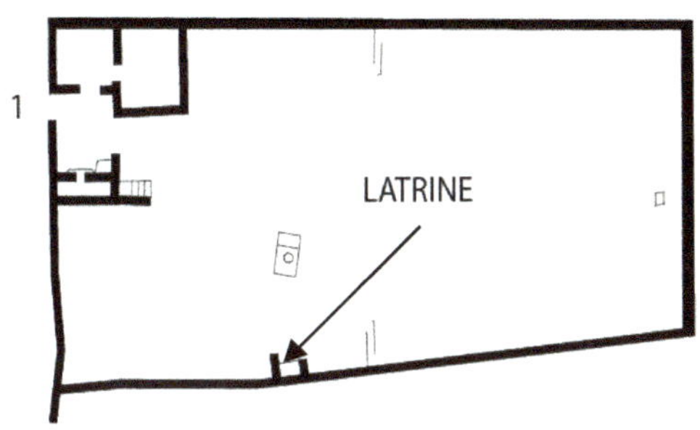

II.8.1

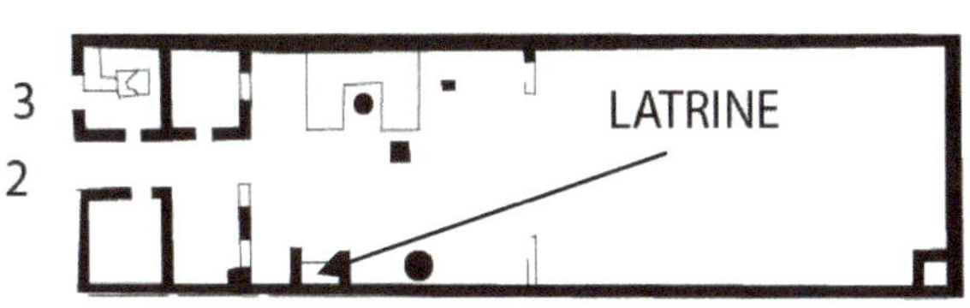

II.8.3

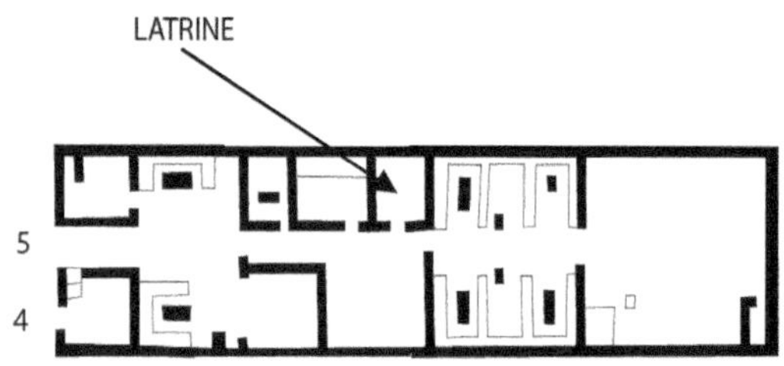

II.8.5

REGIO III

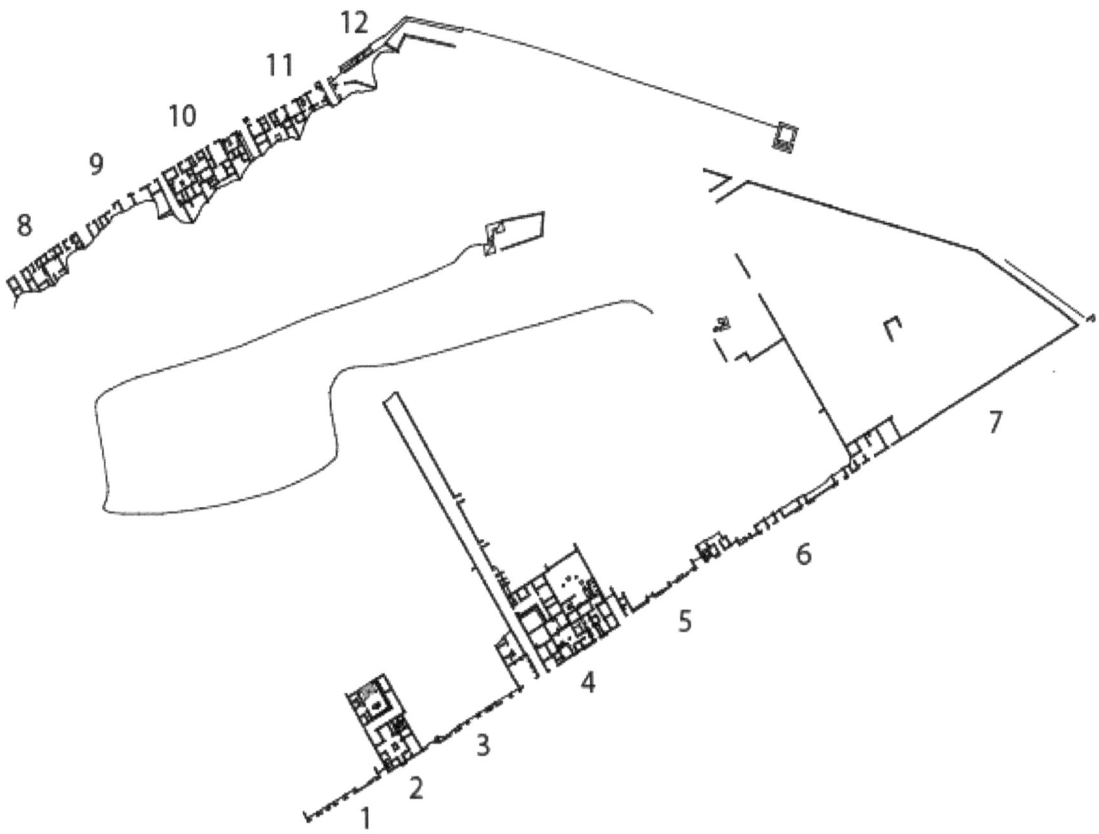

A region which is mainly unexcavated. The occasional down pipe can be seen from the street.

REGIO III.1

III.1.3 Down pipe

REGIO III.4

III.4.2/3 Down pipe in wall

REGIO V

An interesting region, with several unexcavated insulae to the north. A number of upper storey latrines have been identified

REGIO V.1

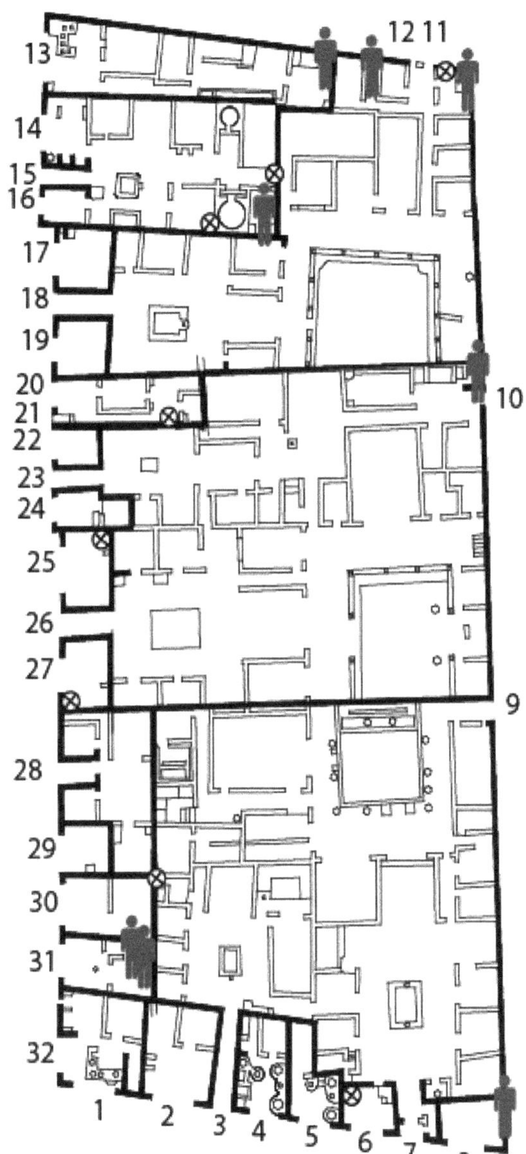

In this insula there are twenty two properties. Thirteen of these are shops and four are workshops. Three large houses take up most of the insula. Two upper storey latrines are visible and the down pipes provide evidence of more upper storey facilities. There are also water pipes and the latrine of the House of Caecilius has a water basin

V. 1. 1, 32 *Thermopolium of Fortunatus*

Latrine under stairs (Eschebach p122)

Nothing

V. 1. 2 *Caupona of Fortunatus*

Latrine in kitchen (Eschebach p122)

Probable latrine

V. 1. 3, 7, 9 *Casa del Torello di Bronzo*

Latrines in 3 and 'grosse Latrine' in 7 (Eschebach p122)

NO ACCESS

V. 1. 4 *Infectorium*

Nothing

V. 1. 5 *Infectorium*

Nothing

V. 1. 6 *Shop*

DOWN PIPE in NW corner

V. 1. 8 *Shop*

UPPER STOREY LATRINE

V. 1. 13 *Popina of Salvius*

Possible latrine at rear of property, room f

V. 1. 14, 15, 16 *Large House with bakery*

V. 1. 17 *Shop*

Nothing

V. 1. 18. 11, 12 *House of the Greek Epigrams (House of L Valerius Flaccus etc)*

Two possible latrines via entrance #11t

LATRINE A in room to left (room t)

LATRINE B in room to right (room U)

V. 1. 19 *Shop*

Evidence of previous excavation??

V. 1. 20, 21 *Shop of Faustus*

V. 1. 23, 25, 26, 10 *House of Caecilius*

DOWN PIPE

Iucundus

Latrine in kitchen 23 (Eschebach p127)

LATRINE with water basin

V. 1. 25 *Shop*

Scar on eastern wall suggesting down pipe

V. 1. 24 *Shop*

Nothing

V. 1. 27 *Shop*

Nothing

V. 1. 28 *Workshop of C Cassius Bassus*

Latrine to left (Eschebach p128)

NO ACCESS

V. 1. 29 *Shop of M Tofelanus Valens*

Nothing

V. 1. 30 *Shop of Canices*

UPPER STOREY LATRINE

DOWN PIPE in NE corner

V. 1. 31 *Shop*

Latrine (Eschebach p129)

LATRINE in nich

V.1.2 Possible Latrine

V.1.6 Down pipe in north west corner

V.1.8 Upper storey latrine

V.1.11 (A)

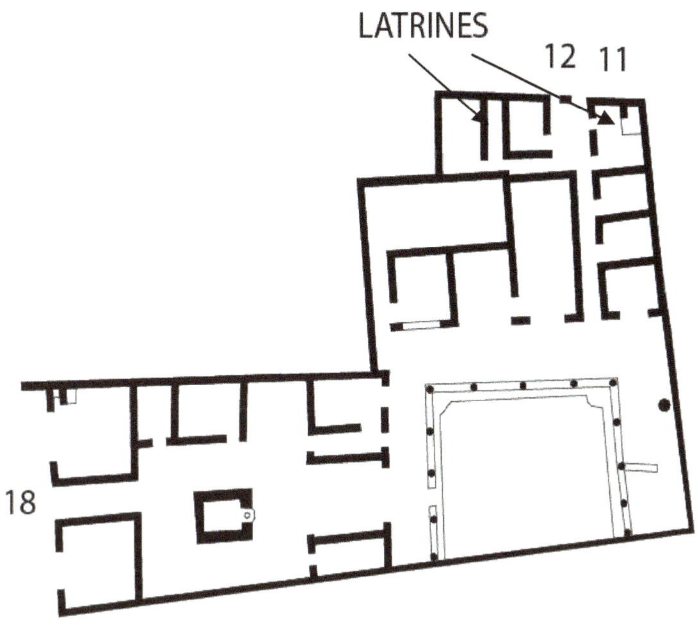

V.1.11,12 Double seat latrine with Down pipe in masonry buttress

V.1.11 (B)

V.1.14-16 ? Latrine

V.1.14-16 Water pipe in east wall

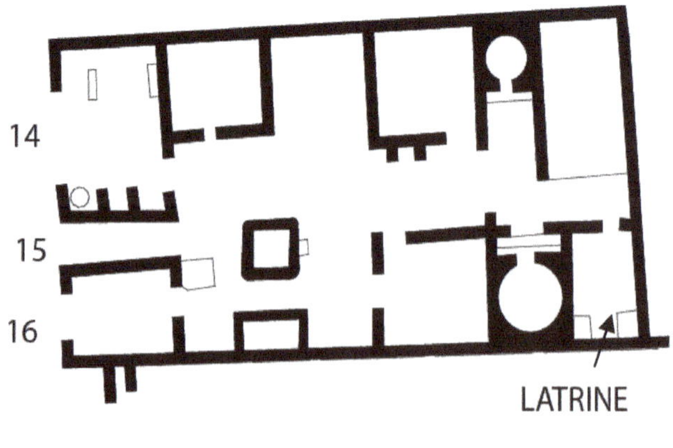

V. 1. 20, 21

V. 1. 23, 25, 26, 10

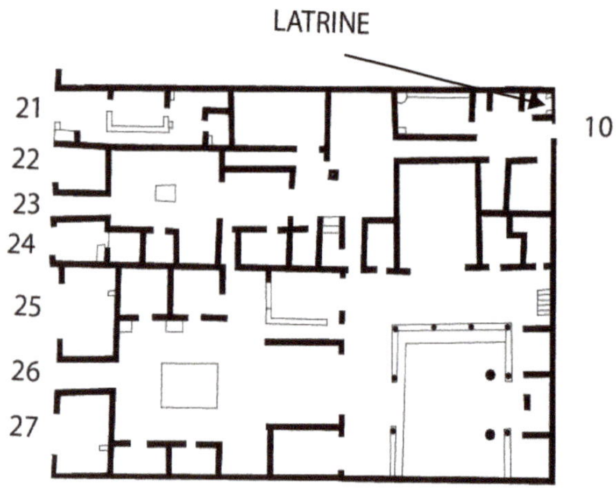

Water Basin by Latrine

V. 1. 25 ? Down Pipe Scar

V.1.27 Down pipe

V. 1. 30

V.1.30 Down pipe in NE corner

V. I. 31

REGIO V.2

In this Insula there are twenty properties, the largest being the House of the Silver Wedding where the decorated latrine room has collapsed. Six other large houses and five smaller ones with eight shops make up the total. Six definite latrines are visible and down pipes in shops #8 and #9 suggest upper storey latrines

V. 2. 1, a *House of Alfius or Sallustius*

LATRINE by entrance to a (Eschebach p129)

DOWN PIPE in outer wall between #1 and #2

V. 2. 2 *Shop*

Nothing

V. 2. 3 *Caupona*

Latrine (Eschebach p130)

Nothing

V. 2. 4 *Casa del Triclinio*

Latrine in kitchen (Eschebach p130)

LATRINE confirmed under stairs

DOWN PIPE

V. 2. 5 *Shop*

Nothing

V. 2. 6 *Shop*

Nothing

V. 2. 7 *Private house with Sheds*

Latrine in kitchen (Eschebach p131)

Nothing

V. 2. 8 *Workshop*

DOWN PIPE

V. 2. 9 *Shop*

DOWN PIPE

V. 2. 10, 11 *House of Paccia*

Latrine in kitchen (Eschebach p132)

Nothing

V. 2. 12 *Shop*

Nothing

V. 2. 13 *Popina*

Latrine in kitchen (Eschebach p132: Packer p35)

Nothing

V. 2. 14 *Popina?*

Nothing

V. 2. 15, 16 *House of Jupiter*

Latrine in 15 with wash basin (Eschebach p133)

Nothing

V. 2. 17-20 *Workshop and thermopolium of Pollia*

Nothing

V. 2. 21 *Shop*

Nothing

V. 2. b, c *Thermopolium*

Latrine in c (Eschebach p134)

Nothing

V. 2. d *House*

Possible in Room L

V. 2. e *Caupona*

Possible under stairs off room b

V. 2. f *Small workshop of Musa*

Nothing

V. 2. g *House of N. Fufidius Successus*

Latrine (Eschebach p135)

Nothing

V. 2. h *House of Aufidia Successa*

LATRINE in room o

V. 2. i, e, 21 *House of Silver Wedding*

Latrine in kitchen in I (Eschebach p136)

LATRINE off room s. Room collapsed

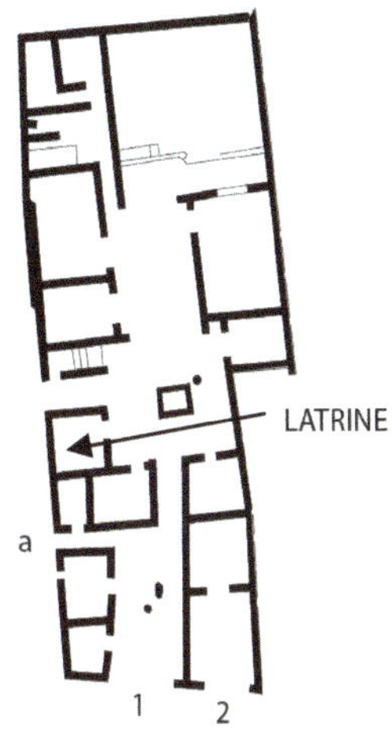

V.2.1, a

V. 2. 4

V.2.2 Down pipe in outer wall by entrance

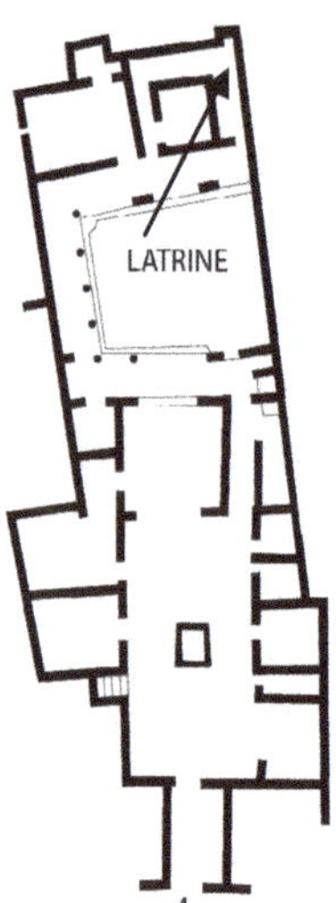

V. 2. 4 ?Latrine under stairs

V. 2. 4 Down Pipe

V. 2. 8 Down pipe

V. 2. 9 Down pipe

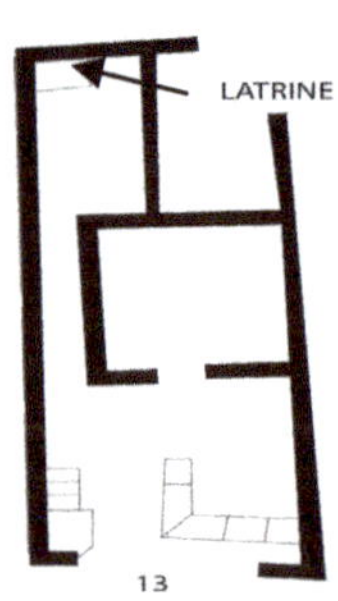

V. 2. 13

V.2.d Latrine in room b

V.2.h Latrine in room o

V.2.i Collapsed latrine room

REGIO V.3

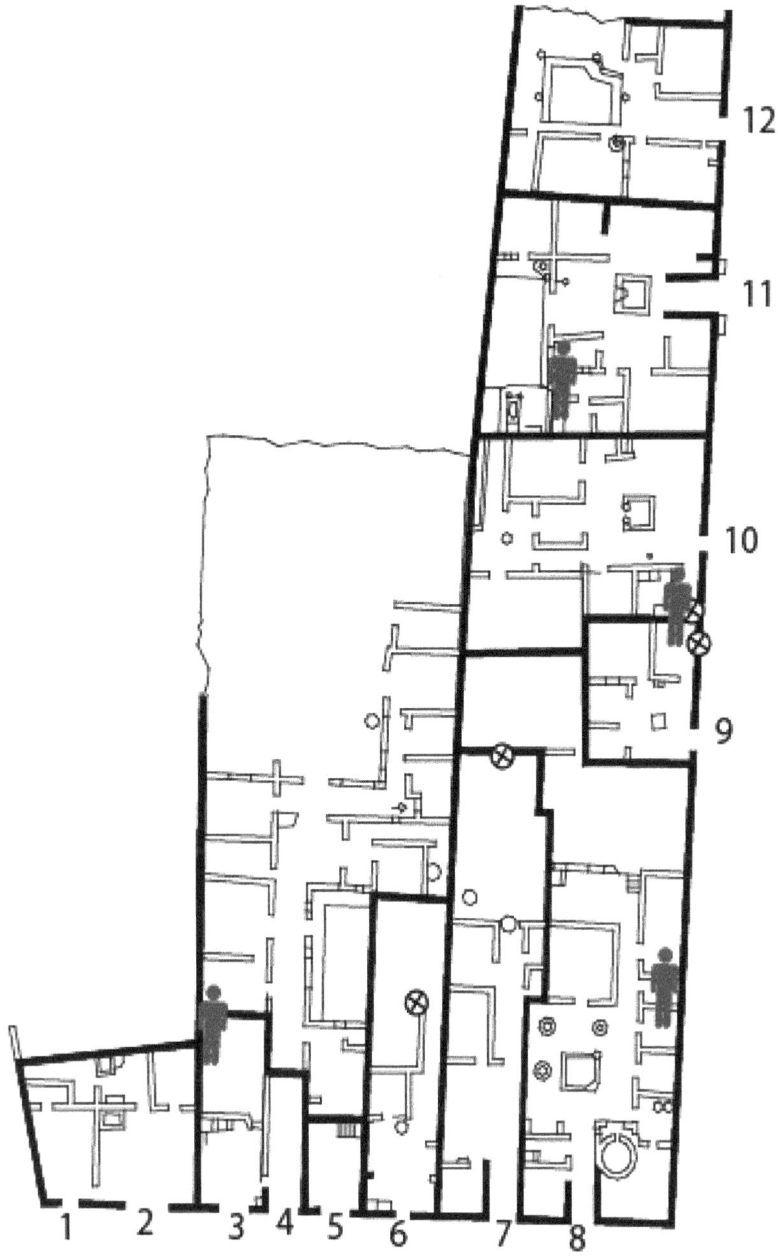

Eight Properties: Four definite latrines.

V. 3. 1, 2 ***Fullonica***

Nothing

V. 3. 3, 4 ***House of Soffitta***

Latrine in kitchen (Eschebach p137)

LATRINE in room c

V. 3. 5 ***Shop***

Nothing

Possible at bottom of stairs

V. 3. 6 ***House of Ceia L F Helpia***

DOWN PIPE In room e

V. 3. 7 ***House of Tiberius Claudius Verus***

Latrine in kitchen (Eschebach p138)

V. 3. 8 ***Bakery***

LATRINE in room 5 (Eschebach p138)

V. 3. 9 ***Small house of Cosmus and Epidia***

DOWN PIPE in outer wall of room B

V. 3. 10 ***Atriumhouse***

LATRINE (Eschebach p139) and DOWN PIPE

V. 3. 11 ***Jewellers House***

LATRINE room E

V. 3. 12 ***House of Samellius Modestus***

Latrine in kitchen (Eschebach p140)

Nothing

DOWN PIPE on north wall. No photograph

V.3.3 Latrine, room c

V. 3. 6 Room e

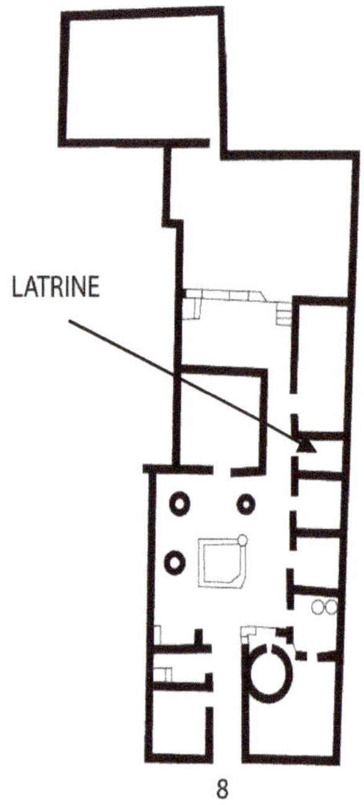

V. 3. 8

V. 3. 10 Latrine & Down pipe

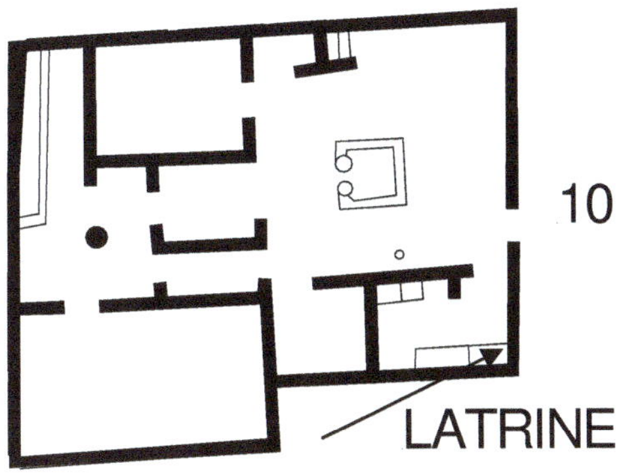

V.3.10 Latrine pipe drain

V.3.11 Latrine before cleaning

V.3.11 Latrine, after cleaning (Wrong board number)

REGIO V.4

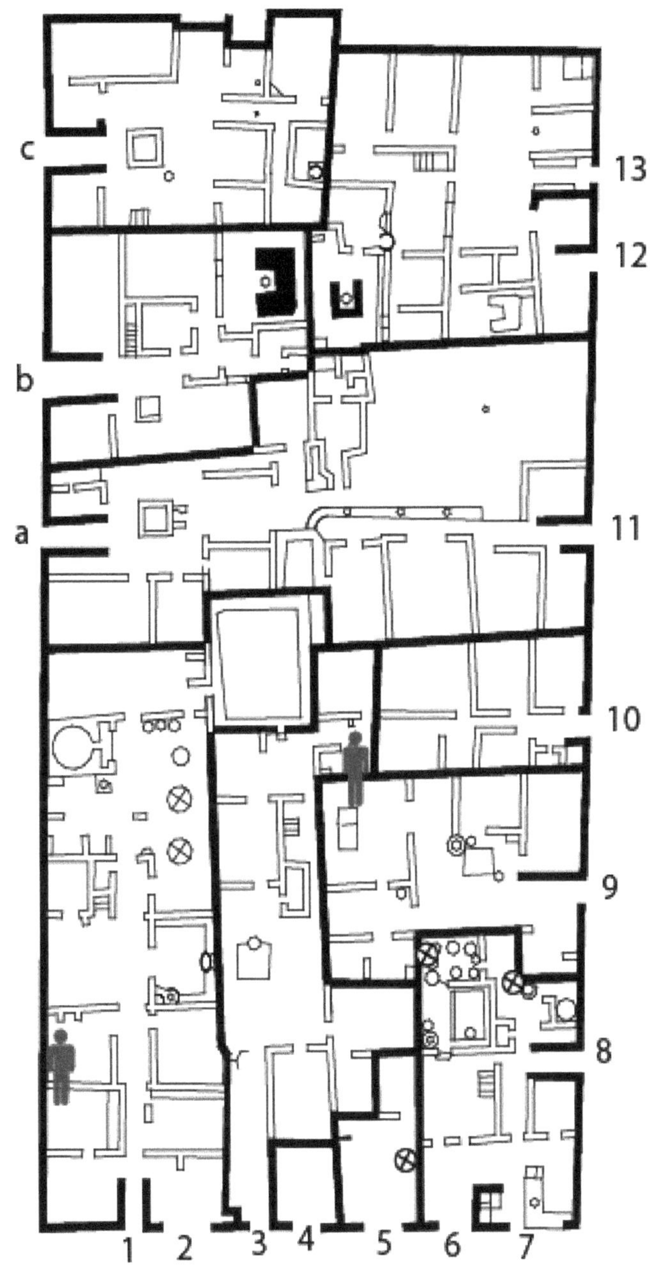

Twelve properties. Two latrines and three down pipes.

V. 4. 1, 2 *Bakery*

LATRINE in room G, in 1(Eschebach p140)

V. 4. 3 *House of a Flamen*

LATRINE in kitchen, room o (Eschebach p141)

V. 4. 4 *Shop*

Nothing

V. 4. 5 *Shop*

DOWN PIPE on western wall

V. 4. 6 *Caupona*

TWO DOWN PIPES

V. 4. 7, 8 *Thermopolium*

Nothing

V. 4. 9 *Workshop*

Nothing

V. 4. 10 *House of M Sittius Potitus*

Latrine (Eschebach p142

Nothing

V. 4. 11, a *House of MM Lucretii, Fronto and Lerus*

NO ACCESS

V. 4. 12, 13 *House of M Fabius Secundus*

NO ACCESS

V. 4. b *Small house*

Latrine in kitchen (Eschebach p144)

Down pipe scar in triclinium area

V. 4. c *House of M Samellius Modestus*

Nothing

V.4.1 Latrine and down pipe

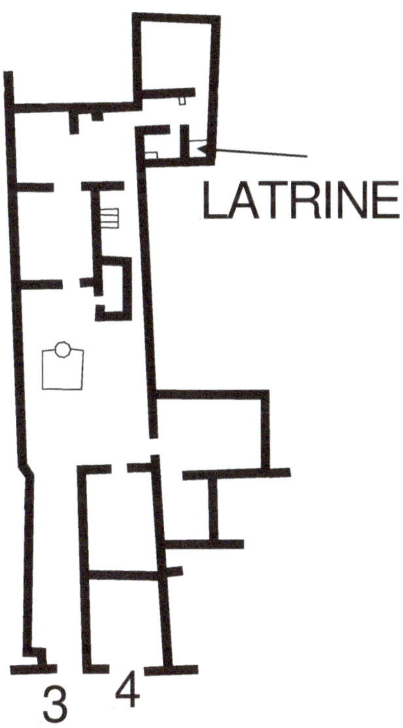

V. 4. 3

V.4.5 Down pipe

V.4.6 Downpipe (i)

V.4.6 Downpipe (ii)

REGIO V.5

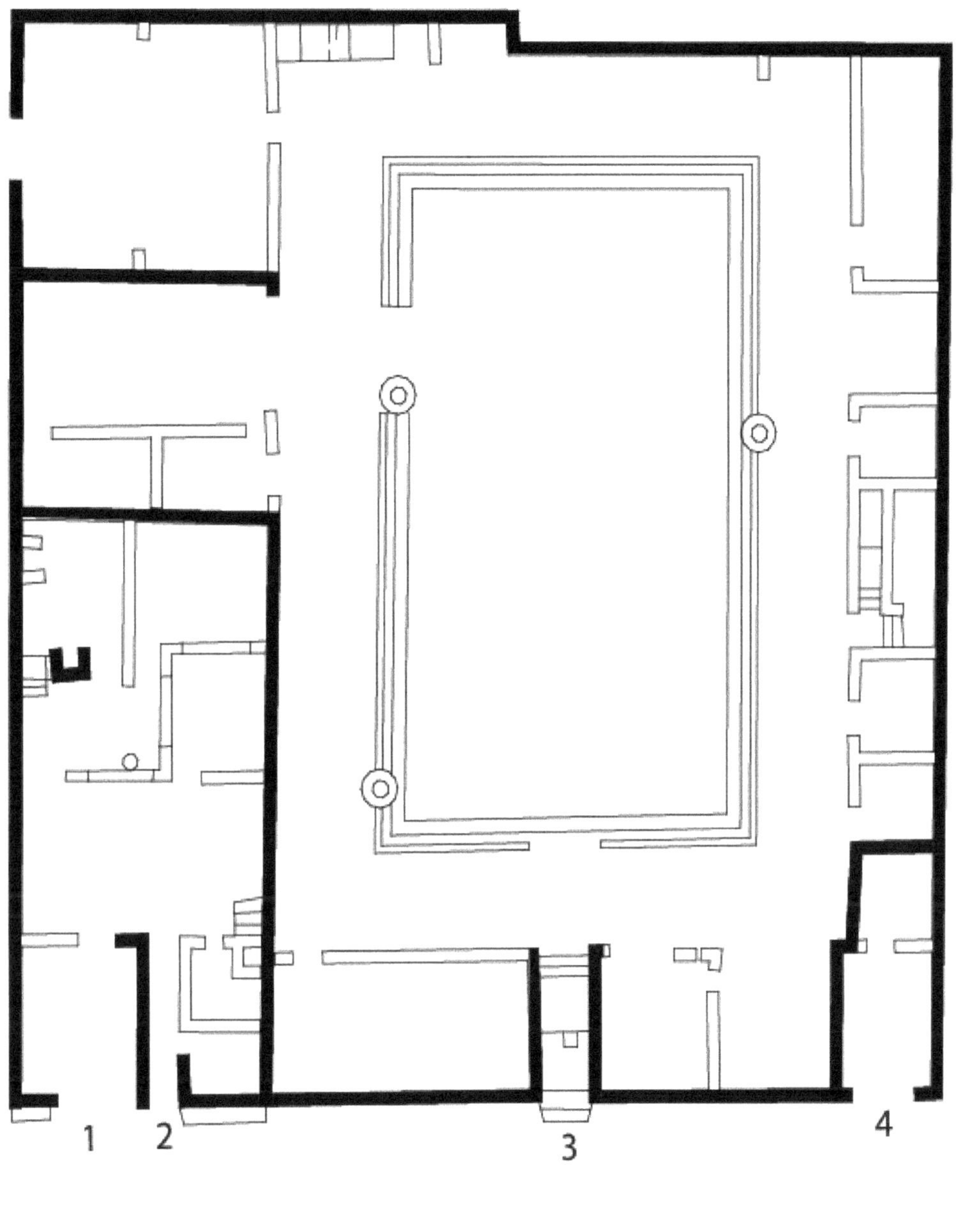

V. 5. 1, 2 House of the doctors of the gladiators

Nothing

V. 5. 3, 4, a House of the Gladiators

Possible latrine in 3 (Eschebach p145)

Nothing

V. 5. 4 Shop

Nothing

REGIO VI

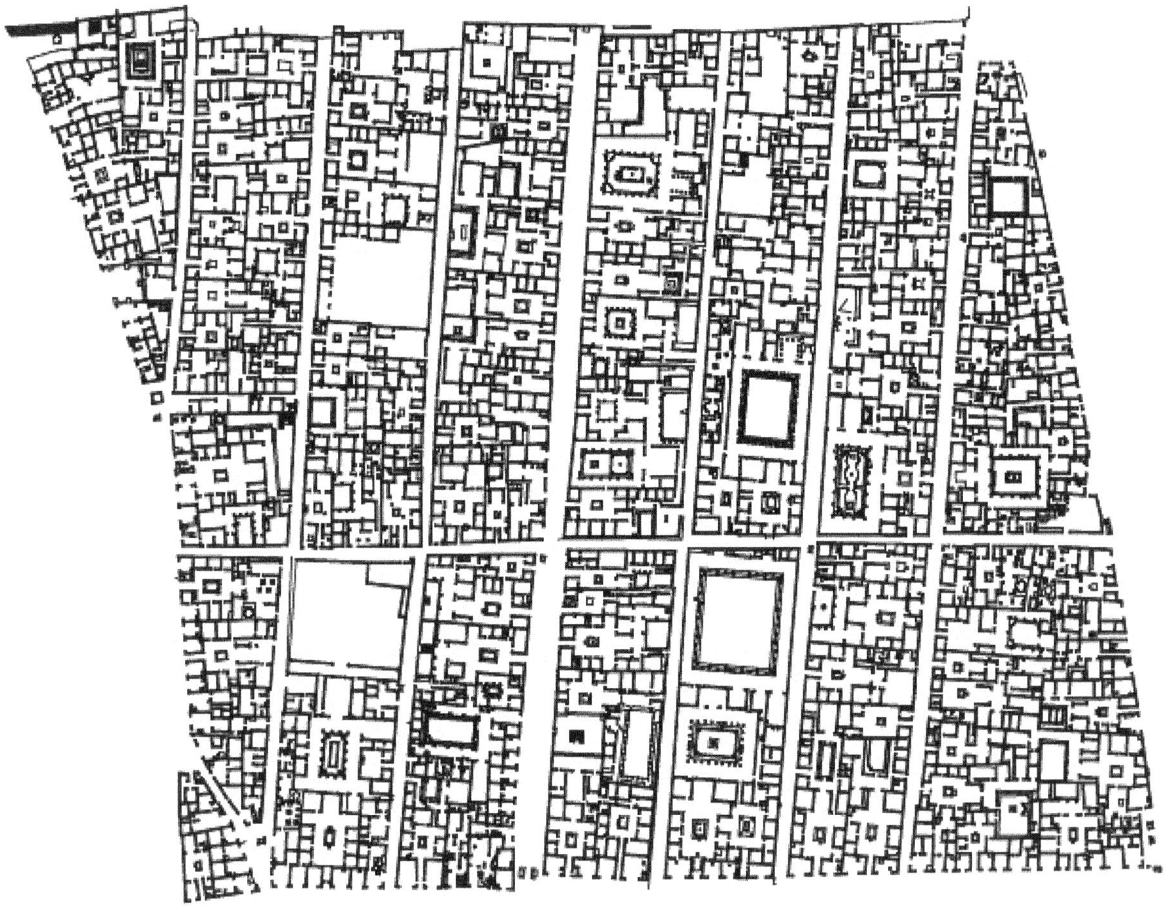

REGIO VI.1

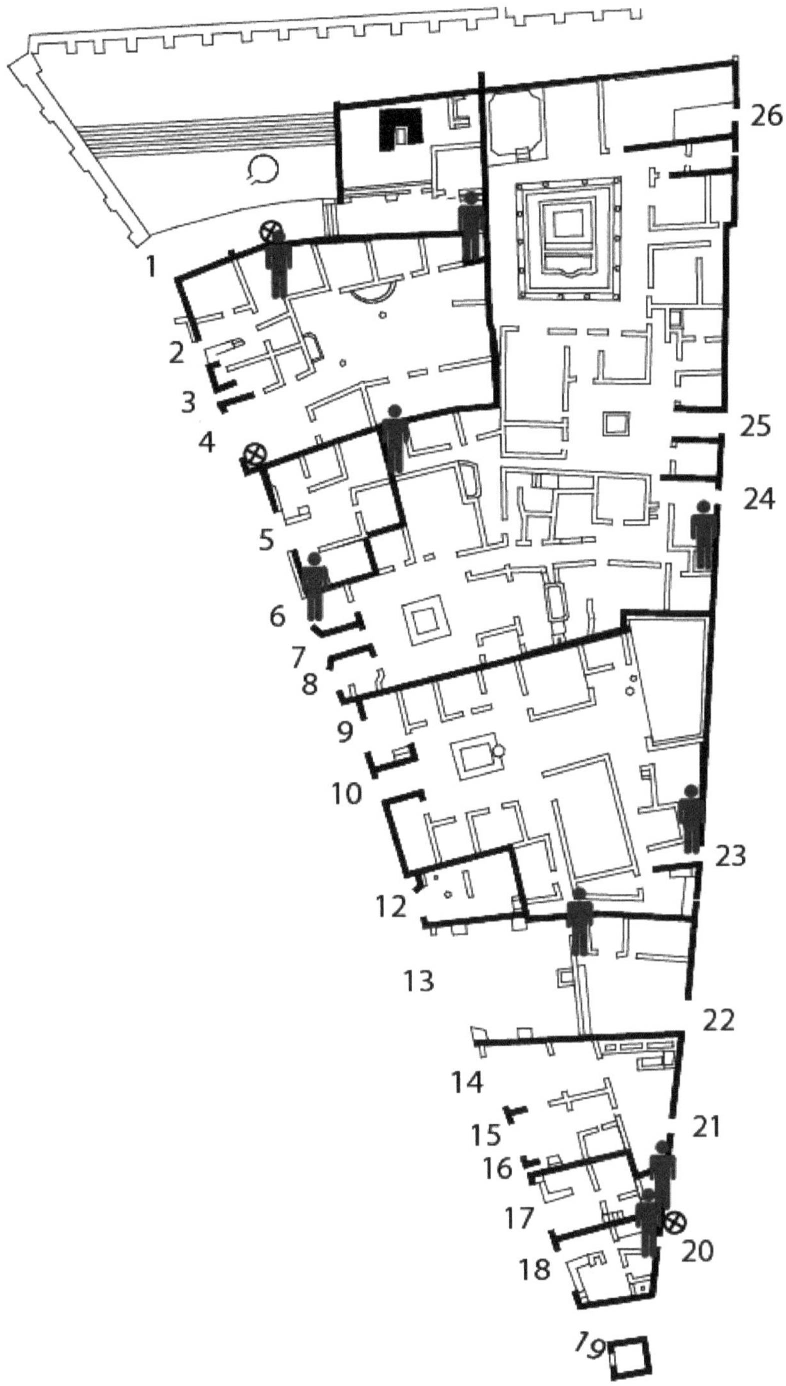

This insula was originally excavated at the end of the eighteenth century. The latrine in VI.1.1 was only visible after excavation although Fiorelli mentions it. The latrines in VI.1.2 and VI.1.24 were covered over by late floors and were discovered by the excavations carried out by the Anglo-American Project in Pompeii

VI.1.1 ***Casa del Triclinio***

DOWN PIPE in buttress of wall

LATRINE and cess pit in VII.1.1 south-east corner of room 4 (Fiorelli p47)

I.1.2,3,4 ***Inn***

DOWN PIPE inside entrance #4 to south

Latrine (Eschebach p151:Fiorelli p47)

Previous latrine in room 4

VI.1.5 ***Popina***

VI.1.6,7,8, 24,25,26 ***Casa delle Vestali***

LATRINE in kitchen area (Fiorelli p48)

Previous Latrine in room 19 covered by later floor

VI.1.9,10,23 ***Casa del Chirurgo***

LATRINE

VI.1.12 Custom House

Latrine (Eschebach p153:Fiorelli p49)

VI.1.13,22 Shrine

LATRINE in room 2 (Fiorelli p49)

VI.1.14.15,16,21 Soap Factory

VI.1.17 ***Bar of Acisculus***

LATRINE in rear room 2 (Eschebach p154:Fiorelli p49)

VI.1.18,20 ***Bar Of Phoebus***

LATRINE in room 3 (Eschebach p154:Fiorelli p49)

DOWN PIPE inserted into wall

Latrine in VI.1.1. Casa del Triclinio

Down pipe in 'Buttress' on ramp of VI.1.1

Latrine in the Inn VI.1.2

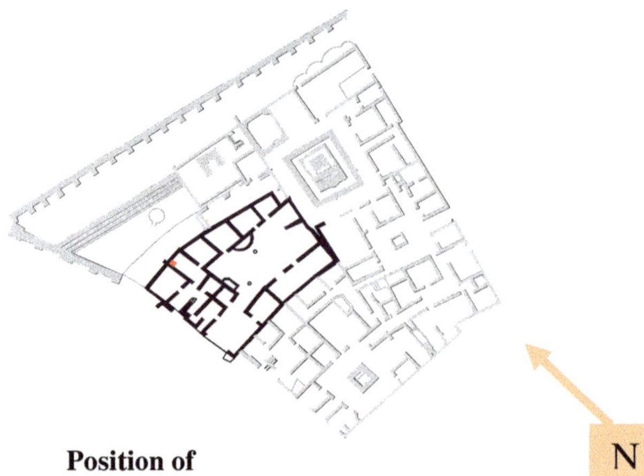

Position of latrine in Inn VI.1.2 Room 5

VI.1.4 Down pipe

VI.1.6,7,8, 24,25,26 *Casa delle Vestali*

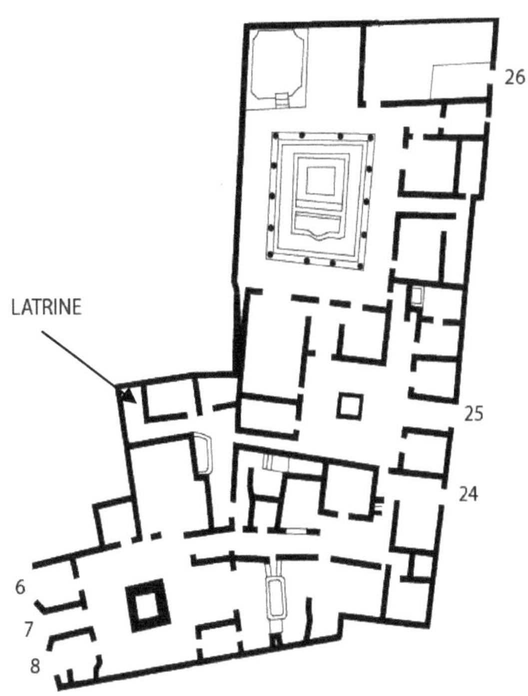

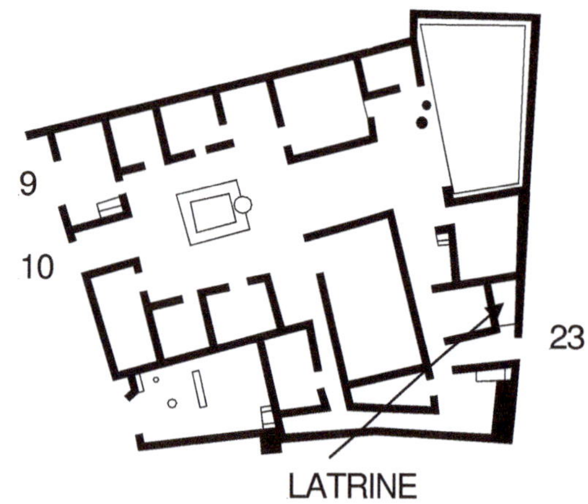

Latrine in Casa del Chirurgo during excavation
(Permission of AAPP)

Drain from latrine of Casa del Chirurgo into cess pit in Vicolo di Narciso

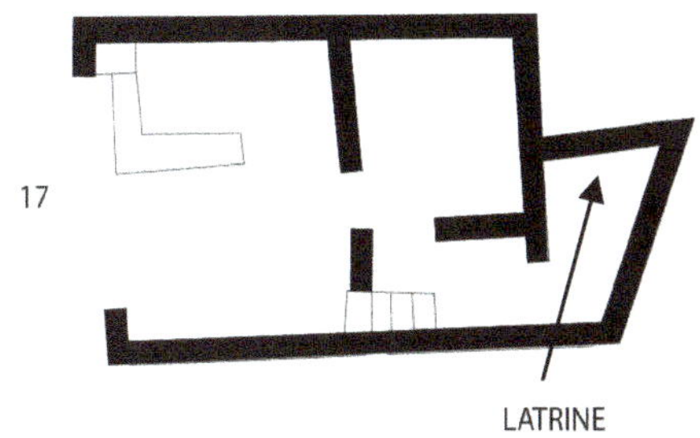

VI.1.17 Latrine in the Bar of Acisculus

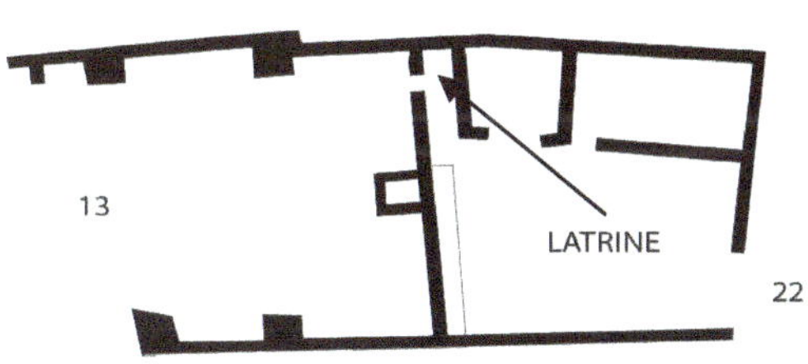

VI.1.22 The latrine 'behind' the shrine

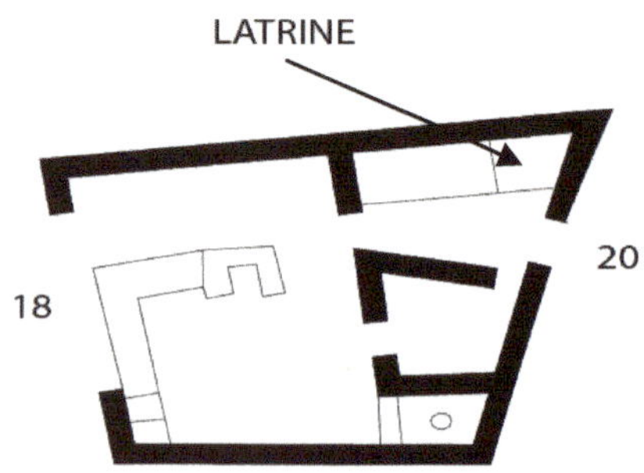

Latrine in Bar of Phoebus VI.1.18,20

Down Pipe running down the outside of the wall of the
Bar of Phoebus in the Vicolo di Narciso

REGIO VI.2

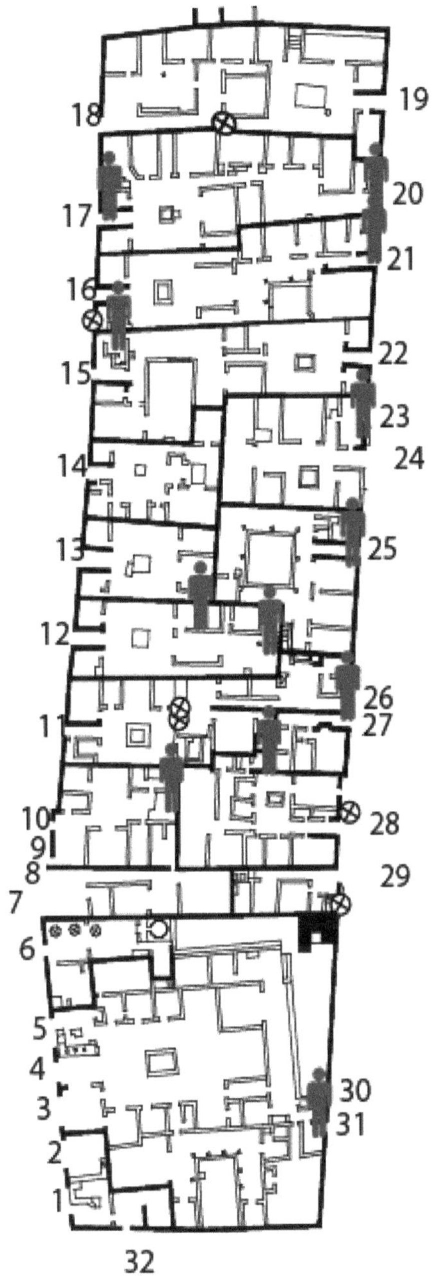

Fiorelli describes five locations for latrines.

There are twenty properties, listed below, with thirteen either definite or possible latrines. Only one down pipe has been found but some narrow water pipes are evident

VI. 2. 1, 32 ***'Thermopolium'***

(Fiorelli p49).

VI. 2. 2 ***'One room property'***

Nothing.

VI. 2. 3, 4, 5, 30, 31 ***Casa di Sallustio***

LATRINE at the rear of the house by entrance #31 in a narrow room. (Eschebach p156:Fiorelli p50)

VI. 2. 6 ***'Bakery'***

(Eschebach p157:Fiorelli 1875:50)

A possible site for this is by the south wall opposite the oven.

VI. 2. 7, 8 ***'Six roomed property'***

Latrine (Eschebach p157)

Nothing

VI. 2. 9, 10 ***'Small house'***

LATRINE in the kitchen area to the north east in which there is a subdivision with the plaster indicating a seat.

(Eschebach p157:Fiorelli p50)

VI. 2. 11 *'Small house'*

(Eschebach p157:Fiorelli p50).

Two WATER PIPES in room 2

VI. 2. 12 ***'Small house'***

Possible in the kitchen area.

VI. 2. 13 'Small house'

Possible at the eastern end of this property in a narrow room (Eschebach p158)

VI. 2. 14 ***Casa delle Amazzoni***

Nothing.

VI. 2. 15, 22 ***Casa di M.Pupius***

Lattrine (ESchebach 158)

Nothing

VI. 2. 16, 21 ***Casa di Narciso***

Possible in a small room 16 with a step up, near to the doorway #16.

At the easterly end of the house, to the north of door #21, room 22 is a more likely place.

LATRINE in the kitchen area in a small room with a window.

VI. 2. 17, 20 ***Casa d'Iside***

Latrine in 17 (Eschebach p159)

Possible by door #20

At the western end of the property there is another small room which appears to be more likely

VI. 2. 18, 19 ***'Hospitium'***

Nothing

VI. 2. 23, 24 ***Casa dei Cosei***

Latrine in 23(Eschebach p160)

Possible just inside the doorway #23 in the kitchen area which has a small wall subdivision

VI. 2. 25 'Courtyard house'

Possible in the kitchen area of this house next to the cooking surface

VI. 2. 26 ***'Workshop'***

LATRINE in a small room, just inside the doorway #26 to the north At the far end is the outline, in the masonry, of the wooden seat of the latrine.

VI. 2. 27 ***'Small house'***

Latrine (Eschebach p161)

Possible in a small room with a plaster niche let into the wall, at the western end of this property

VI. 2. 28 *'Hospitium'*

Latrine (Eschebach p161)

Nothing.

VI. 2. 29 *'Small workshop'*

DOWN PIPE

VI. 2. 3, 4, 5, 30, 31

VI. 2. 6

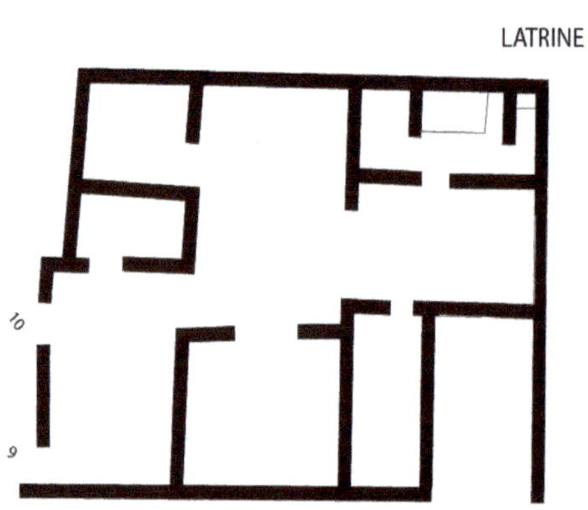

VI.2.10

VI.2.12

VI.2.12

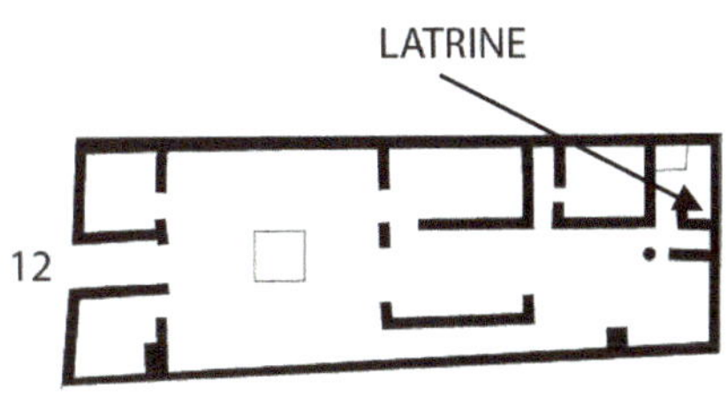

VI.2.11

VI.2.13

VI. 2. 16. 21

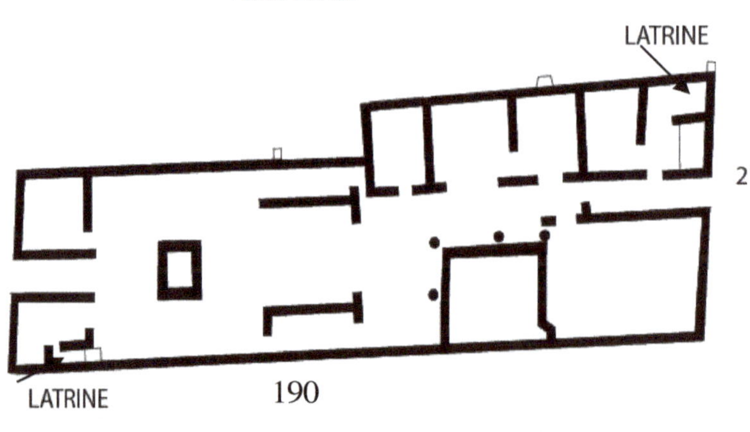

VI. 2. 16, 21 kitchen area

VI. 2. 17, 20 by door #20

VI. 2. 23, 24

VI. 2. 26

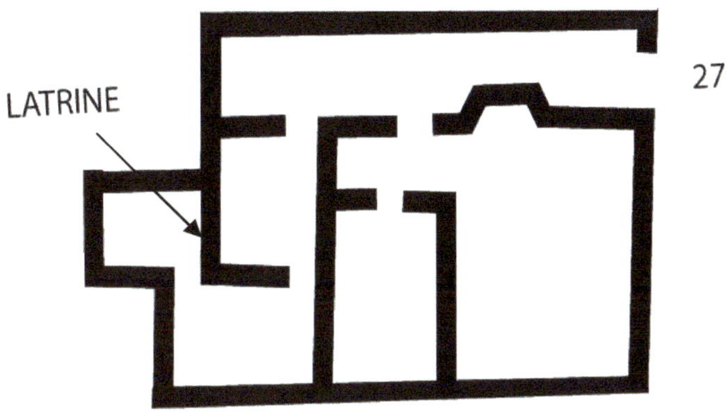

VI. 2. 27

VI. 2. 28 Down pipe to north of threshold of #28

VI.2.29 Down pipe

REGIO VI.3

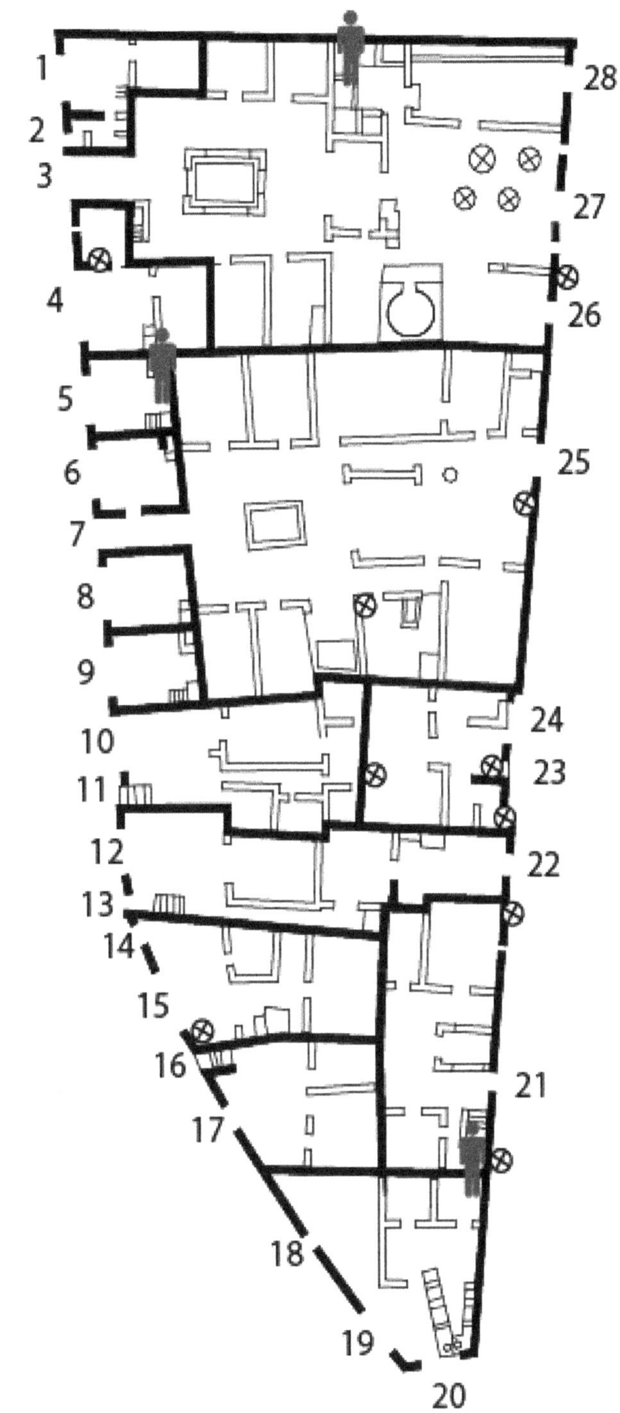

Insula 3 has fourteen properties. Of these only two are of any size. There are a number of downpipes suggesting that that there were latrines in an upper storey.

VI. 3. 1, 2

A small property. Nothing found

VI. 3. 3, 26, 27, 28

LATRINE in the kitchen area (Eschebach p162)

DOWN PIPE at the rear of this property, of external diameter 31cm,in the wall between entrances 26 and 27

VI. 3. 4

Nothing

VI. 3. 5 A small room, possibly the latrine, filled with rubble. (?Eschebach p 162)

Thèdenat (1910:97) notes an upper storey pipe, no evidence for which was found

VI. 3. 6, 7, 25

Latrine in 6 (Eschebach p163)

Thèdenat (1910:96) notes that the latrine in this property was connected to a very deep ditch.

A DOWN PIPE can be seen, looking to the east from the entrance #7, in the eastern wall just to the south of entrance #25

VI. 3. 8

Nothing

VI. 3. 9

Latrine (Eschebach p163)

This property is a small shop in which there is no evidence of a latrine

DOWN PIPE in the wall, between the entrances 9 and 10.

VI. 3. 10, 11

Latrine (Eschebach p164)

Nothing

VI. 3. 12, 13, 22

Nothing

VI. 3. 14, 15

Latrine (Eschebach p164)

DOWNPIPE of diameter 20cm in the southern wall

VI. 3. 16, 17 There is a scar in the wall between entrances 16 and 17 which may have held a down pipe. This appears to discharge onto the road

VI. 3. 18, 19, 20

Latrine in 20 (Eschebach p165)

Nothing

VI. 3. 21

Latrine in kitchen (Eschebach p166)

VI. 3. 23, 24

Latrine in 24 (Eschebach p166)

DOWN PIPE of narrow bore. External diameter 11cm. This might be for rainwater.

VI.3.5

VI.3. 6,7,25 Down pipe in east wall

Down pipe between VI.3.9 and VI.3.10

VI. 3. 14, 15 Down pipe

Scar of down pipe in wall between

VI. 3. 16 and VI. 3. 17

VI. 3. 23 Room 2

Narrow bore down pipe between entrances VI. 3. 23 and VI. 3. 24

VI.3.26 Down pipe in wall to north of entrance

REGIO VI.4

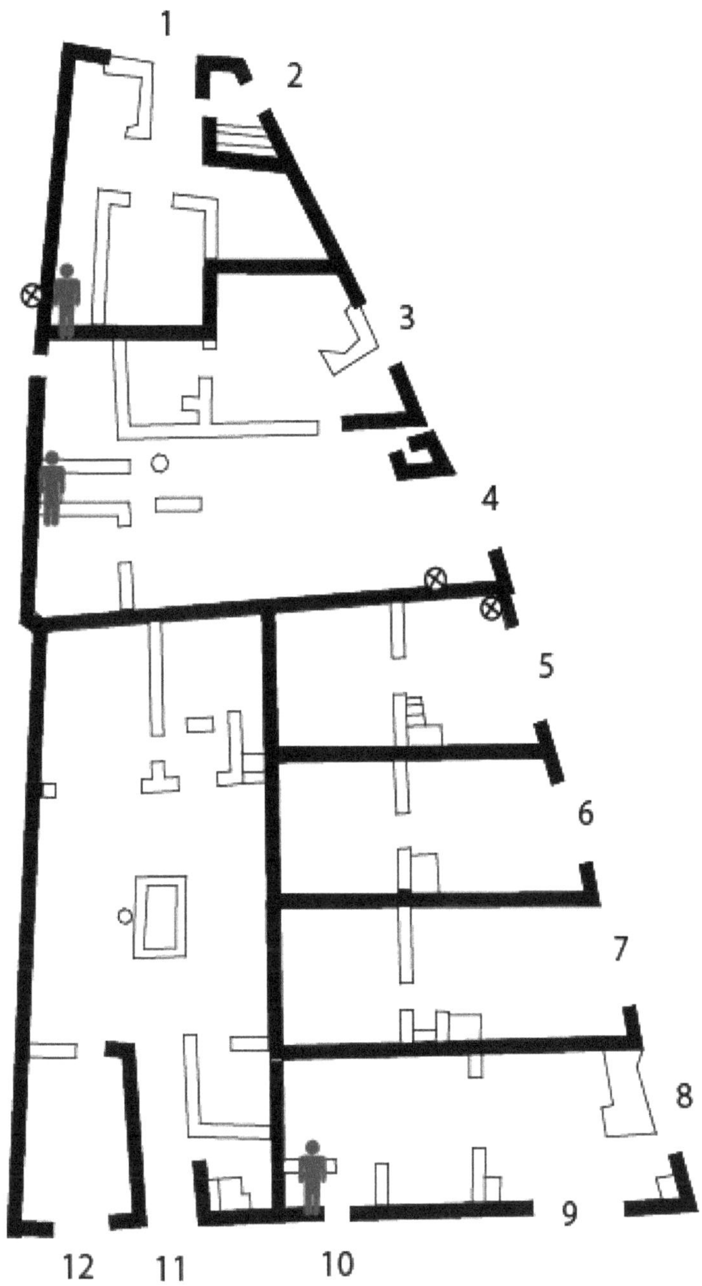

There are eight properties in VI.4.

The diameter of one of the down pipes suggests at least one latrine in the upper storey

VI. 4. 1, 2

Latrine in 1 (Eschebach p166: Fiorelli p53)

DOWN PIPE of external diameter 22cm in a long narrow room to the southwest of the property. This is just visible at the furthest corner on the right of the photograph.

There was probably a latrine at the far end of this room

VI. 4. 3, 4 *Hospitium*

Latrine in 4 (Eschebach p167: Fiorelli p53)

DOWN PIPE set into the southern wall, external diameter 13cm.

At the rear of the property there is a small room the northern wall of which is no longer standing, despite being shown on the plan. This room has a sloping tile feature. There is a window above where the latrine appears to be.

VI. 4. 5 *Shop*

DOWN PIPE in North west corner room 1

VI. 4. 6 *Shop*

Latrine (Eschebach p167)

Nothing

VI. 4. 7 *Large shop*

Nothing

VI. 4. 8-10 Large shop

Latrine in 9 (Eschebach p168)

Possibly a latrine in room 2 in the southwest corner of the property

VI. 4. 10, 11

Latrine in 11 (Eschebach p168)

Nothing

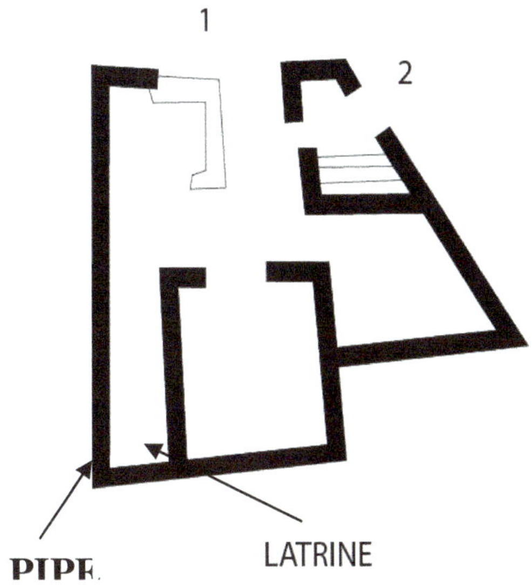

VI. 4. 1, 2

VI. 4. 1, 2 Downpipe in outer wall

VI. 4. 4 Downpipe

VI. 4. 4 Latrine

VI. 4. 5 Collapsed floor showing down pipe

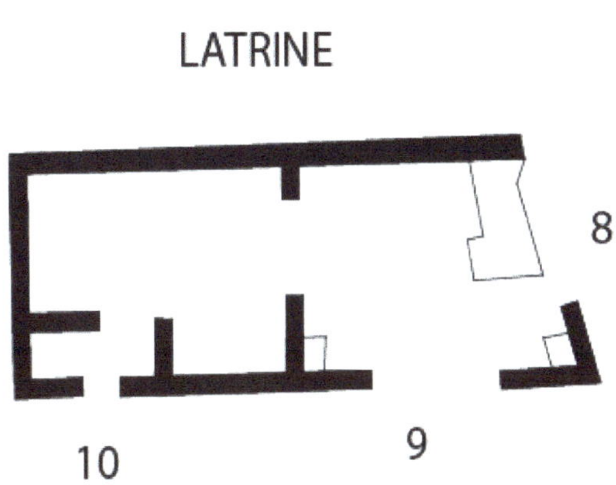

VI. 4. 8, 9, 10 Latrine

REGIO VI.5

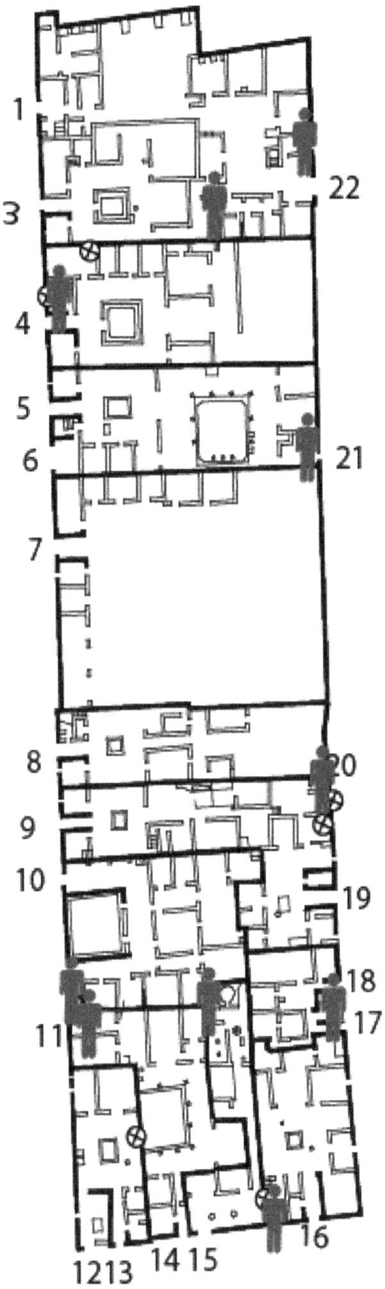

There are twelve properties in *Insula* 5. Most of these are houses but there is one which is mainly a garden and one industrial premise. There are a considerable number of latrines at ground floor level. Nevertheless there is evidence of at least one upper storey latrine

VI. 5. 1, 3, 22 ***Casa di Nettuno***

The room in the north west corner of the house which unfortunately is filled with rubble is a possibility. Also possible is near the internal staircase (VI.5.1,3,22A).The narrow room to the north just inside entrance #22 is more likely (VI.5.1,3,22B).

Latrine in kitchen in 3 (Eschebach p169)

VI. 5. 4 ***'Atrium house'***

LATRINE constructed of brick. It may be a late addition to the house (Eschebach p169)

DOWN PIPE, external diameter 22cm, inserted into the wall behind the latrine is visible on the western wall of the house to the north of entrance #4.

WATER PIPE in room 11

VI. 5. 5, 6 ***Casa delle quattro colonne***

Possible just inside entrance #21 (Fiorelli p55)

VI. 5. 7 ***Garden of M. Valerius Abinnericus***

Nothing

VI. 5. 8, 20 ***'Narrow house'***

LATRINE to the north of entrance #8 in a service room. (Eschebach p170)

VI. 5. 9, 19 ***Casa dei Fiori***

LATRINE at the east end of the kitchen area. (Fiorelli p55)

DOWN PIPE, external diameter 10cm, in the outside eastern wall of the kitchen.

DOWN PIPE of external diameter 19cm in the room to the south of the kitchen.

VI. 5. 10 ***'Small house'***

Possible in the kitchen area of this property

VI. 5. 11, 14 ***'Moderately unregal house'***

LATRINE in the north-west corner.

VI. 5. 12, 13 ***'Thermopolium'***

Latrine in kitchen in 13(Eschebach p171)

WATER PIPE in east wall of room 2

VI. 5. 15 ***'Bakery'***

DOWN PIPE in room 1. (Fiorelli p55)

VI. 5. 16 ***Casa di Faventinus***

LATRINE in this room at the south of the property. (Fiorelli p55)

VI. 5. 17, 18 ***Casa della Colonna Etrusca***

Latrine in kitchen in 17 (Eschebach p172)

A small room next to the kitchen. The eastern wall appears to have slots for the seat.

VI. 5. 1, 3, 22A

VI. 5. 1, 3, 22B

Downpipe between #4 and #3

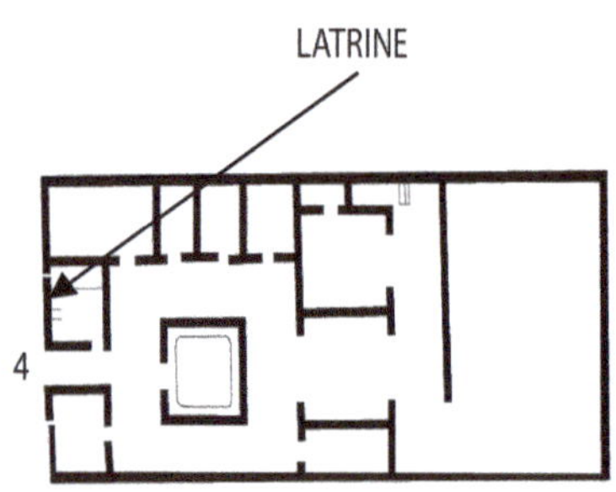

VI. 5. 4

VI. 5. 5, 6

Narrow down pipe VI. 5. 19

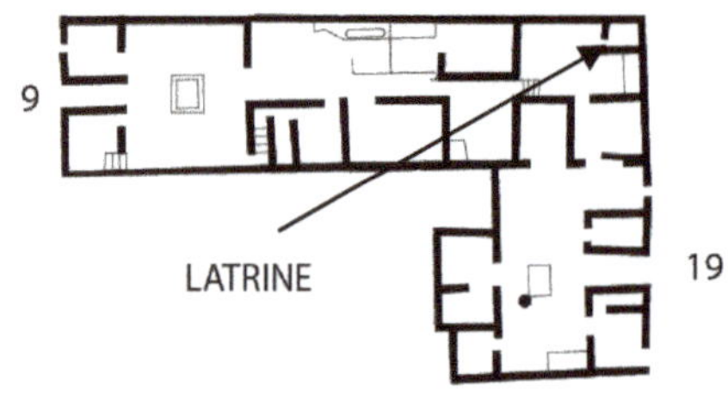

VI. 5. 9, 19

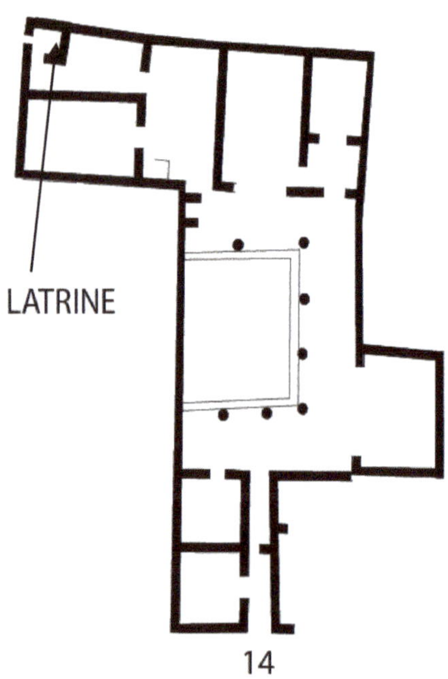

VI. 5. 14

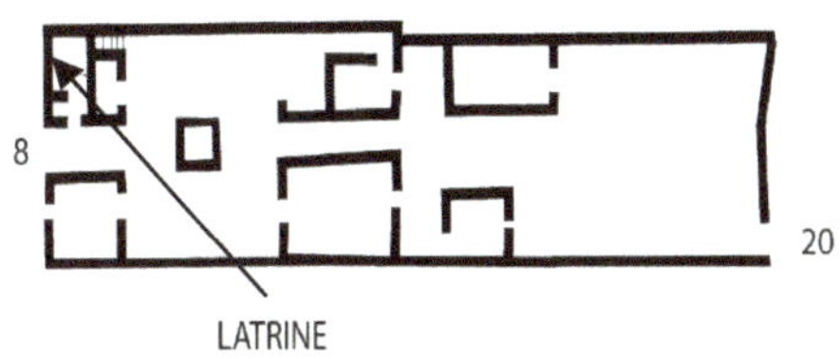

VI. 5. 8. 20

Large diameter down pipe inside wall to
South of Kitchen of VI.5 9,19

VI. 5. 10

VI. 5. 15 Room 1: Down pipe

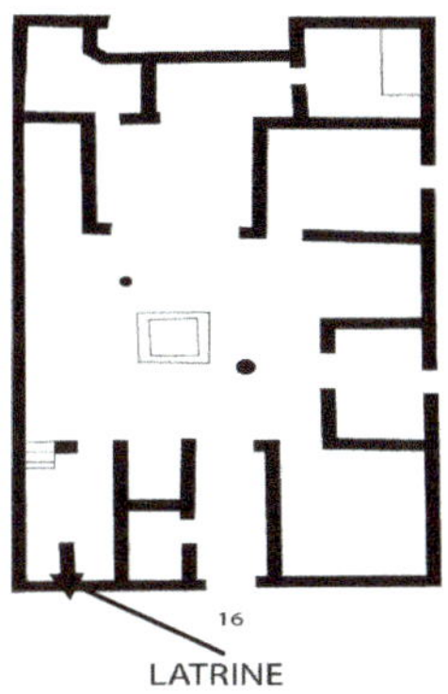

VI. 5. 15

VI. 5. 16

VI. 5. 17, 18

VI.5.18:Down Pipe in outer wall

REGIO VI.6

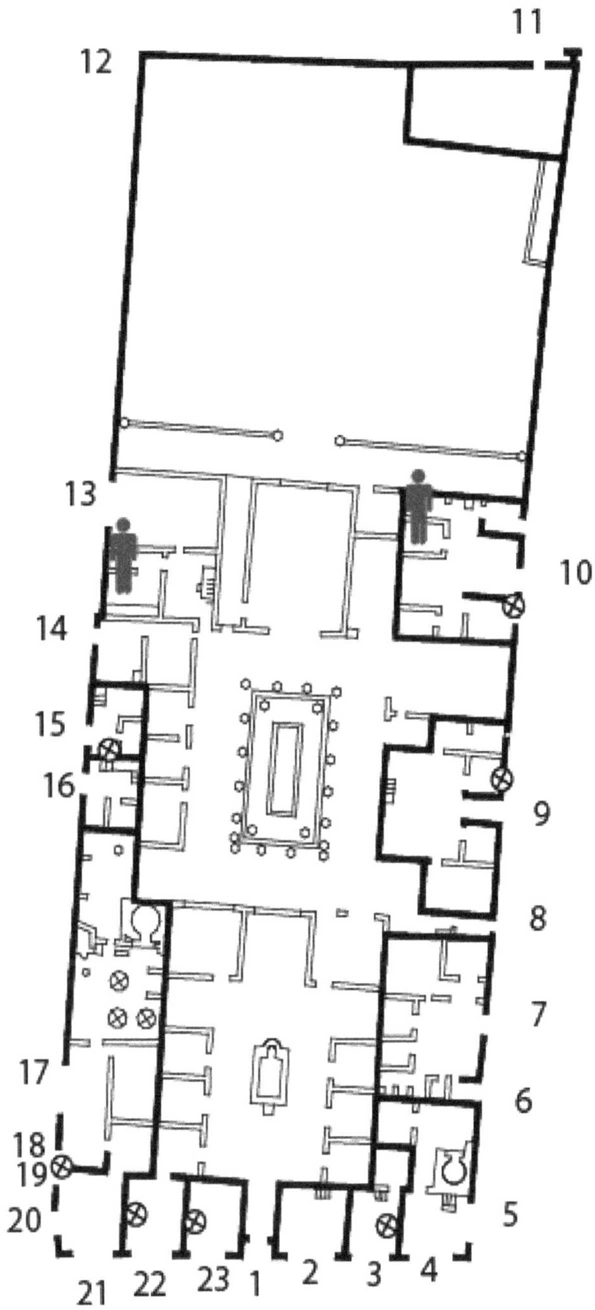

A major proportion of this insula is occupied by the Casa di Pansa and its garden. The house itself has a latrine in the kitchen area to the north-west of the property. Around the front of the house and its sides are small properties, one of which was a bakery. These do not appear to have been serviced with latrines. However there are a number of down-pipes which are of a large diameter indicating that the latrines were in the upper storey.

VI. 6. 1, 8, 12, 13, 22 The Casa di Pansa

LATRINE in the kitchen area at the north west of the property. (Eschebach p173:Fiorelli p55)

VI. 6. 2 Shop

Nothing

VI. 6. 3 Shop

No evidence of a latrine

DOWNPIPE in easterly wall, external diameter 23cm.

VI. 6. 4, 5 Small Bakery

Nothing

VI. 6. 6, 7 Rented House

Latrine (Eschebach p174)

Nothing.

VI. 6. 9 Rented House

DOWNPIPE in the eastern wall. The room behind has no obvious indicators for a latrine.

VI. 6. 10 Rented House

The north west room of this small property seems the most likely place for the latrine (Eschebach p175:Fiorelli p56)

VI. 6. 11 Courtyard

Nothing

VI. 6. 14 Small Workshop

Latrine (Eschebach p175:Fiorelli p56)

Nothing

VI. 6. 15 Small Workshop

Nothing

VI. 6. 16 Small Workshop

Nothing

VI. 6. 17, 18, 19, 20, 21 Rented Bakery

Nothing inside the property. There is a downpipe scar in the outer western wall of diameter 21cm

VI. 6. 22 There is nothing in this property except a DOWNPIPE in the western wall of external diameter 23cm

VI.6.23

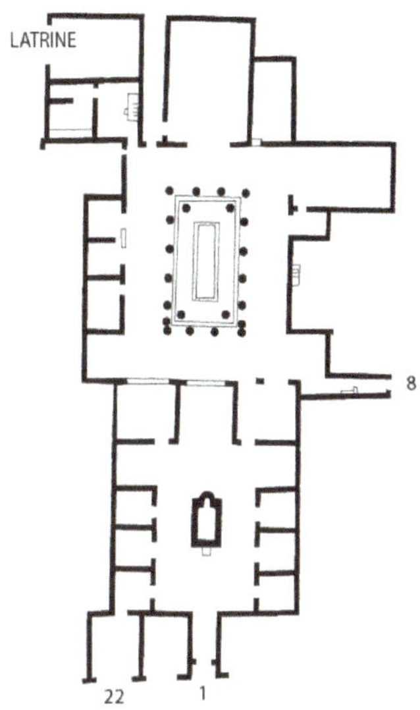

VI. 6. 1

VI. 6. 3 Downpipe

VI. 6. 9

VI. 6. 10

VI.6.10 Down pipe

VI. 6. 15

VI. 6. 19 room 4

VI. 6. 22 Downpipe

REGIO VI.7

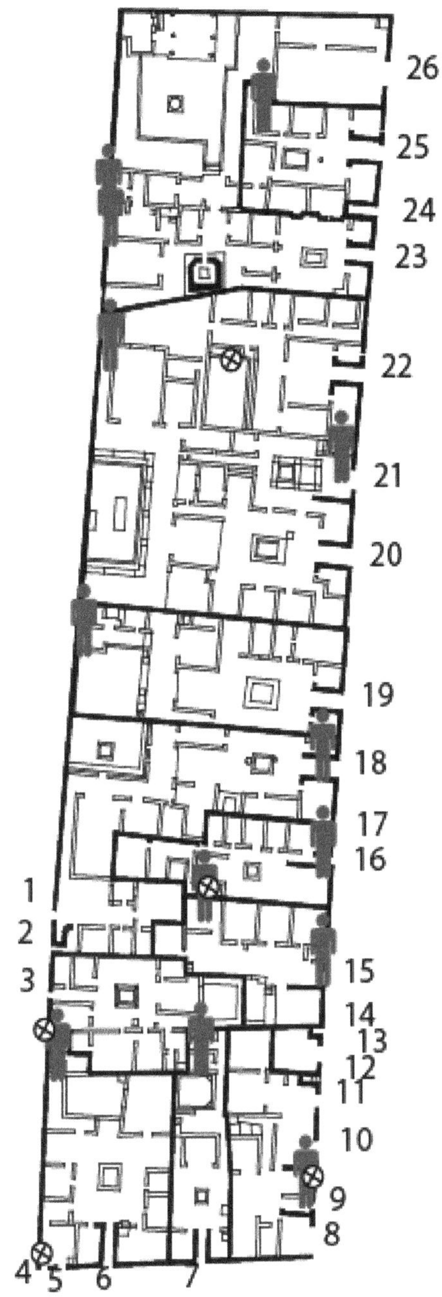

There are fourteen properties and a staircase in this *insula.* Fiorelli mentions three latrines.

As can be seen from the plan most of the latrines are in rooms adjacent to the street

VI.7.1,2,18 Casa di Ione

Possible just inside the entrance #2. (Eschebach in 1, under the stairs p179: Fiorelli p57). Possible to the north of the entrance to #18.

VI.7.3 Casa con atrio tetrastilo

Possible in the kitchen area (Eschebach p 177)

VI.7.4-6 *Casa di Ercole*

LATRINE room at the north-western corner of the property. (Eschebach latrine in 6 p177)

DOWN PIPE to left inside entrance #4

VI.7.7 *'Private house'*

Nothing.

VI.7.8-12 *Casa del Tullius*

Possible in small room under the staircase (Eschebach latrine in 9 p178)

DOWN PIPE outer wall between #9 and #10

VI.7.13,14 *'Shop'*

Nothing.

VI.7.15 *Casa del Duca Aumale*

LATRINE in small room to the right of entrance #15. (Eschebach p179: Fiorelli p58)

VI.7.16,17 *Casa di Mercurio*

Two possible sites for latrines and also a DOWNPIPE of large diameter.

Latrine (Eschebach in 16 p179)

VI.7.19 Casa di Inaco ed Io

LATRINE set into a plastered niche, in the north-western area of this house. (Eschebach p180)

VI.7.20,21,22 Casa d'ell Argentaria

A very large property with two possible latrines and one definite.

The first is to the north of entrance #21.

The second is to the north of entrance #22, accessed by a tortuous route to a narrow room running northwards

LATRINE in a dark vaulted room at the end of a long corridor in the north western part of the property.

(Eschebach two latrines p180)

WATER PIPE in the garden area of #22

VI.7.23 Casa di Apollo

LATRINE in the western side of this property The most southern of these is in the kitchen. It has a raised step and a sloping floor with the escape drain to the rear (Fiorelli p59: Eschebach p181) Beyond the wall to the north of this latrine is a second LATRINE.

VI.7.24 *Staircase*

VI.7.25 *'Small workshop'*

LATRINE in the kitchen of this property. (Eschebach p182)

VI.7.26 *'Shop'*

VI.7.3

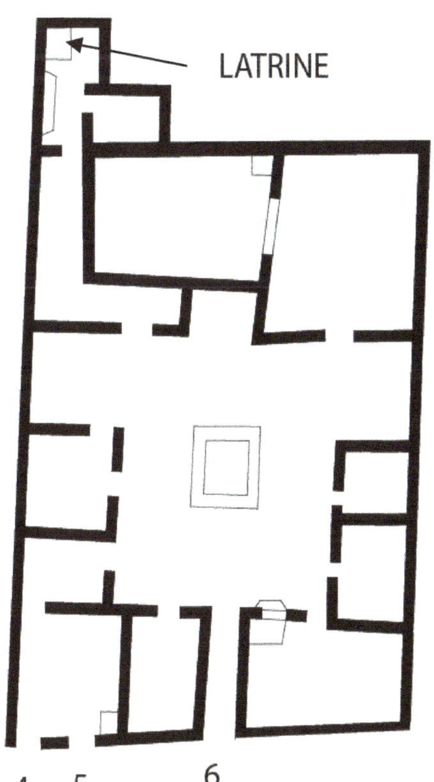

VI.7.4 Down pipe

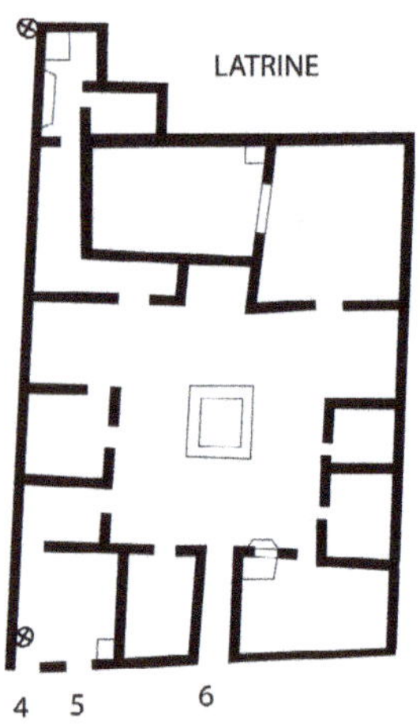

VI.7.4,5,6

VI.7.8-12

VI.7.9,10 Down pipe

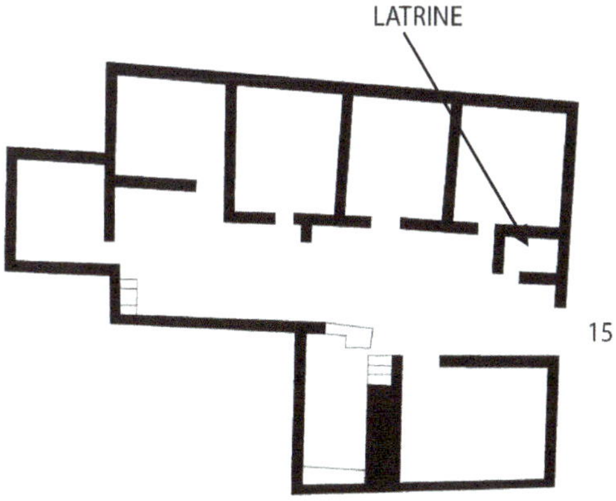

VI.7.15

VI.7.16,17 Down pipe

VI.7.16.17 by staircase

VI.7.18

VI.7.19 Niche latrine

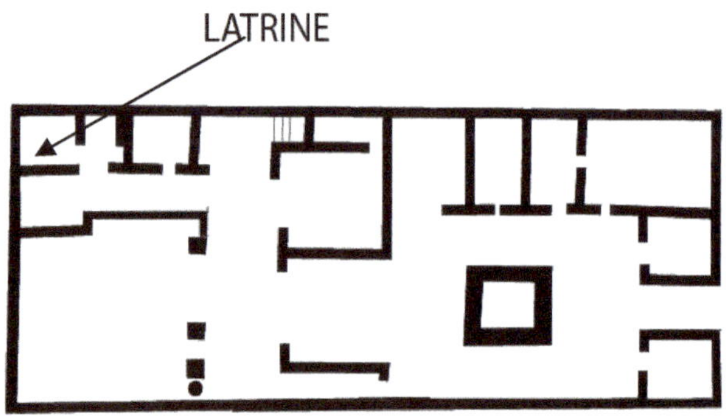

VI.7.20.21.22 Room to north of #21

VI.7.20,21,22 Vaulted room

VI.7.20, 21, 22 Vaulted room after cleaning

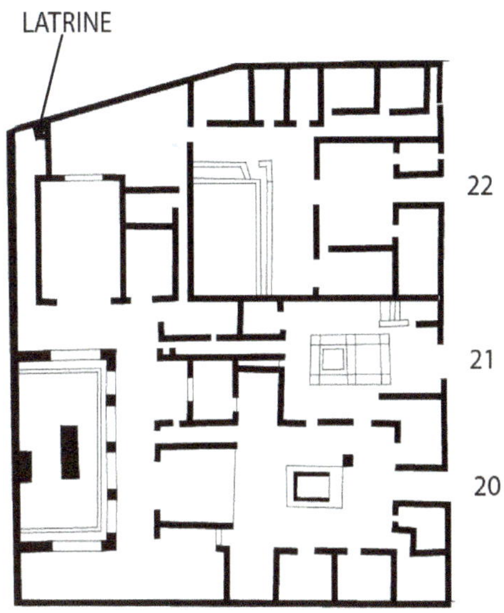

VI.7.20,21,22 Room to north of #22

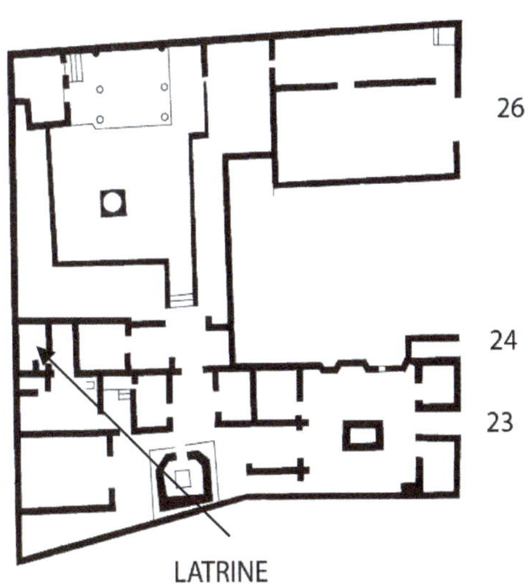

VI.7.23

VI.7.22

VI.7.23 B Possible latrine

VI.7.25

REGIO VI.8

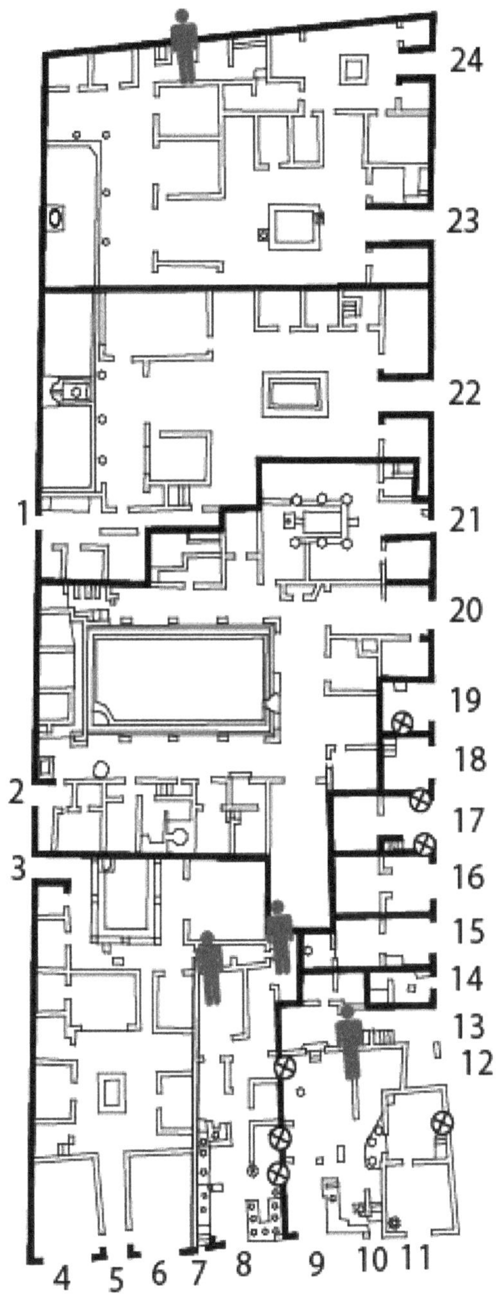

Insula 8 of Regio VI is made up of four large houses, a couple of smaller ones and some shops. No latrines in the shops but one in almost all of the large houses. Down pipes are found in the south eastern section of the *insula* where the shops and bars are.

I.8.1.22 *'Cella vinaria'*

Nothing.

VI.8.2,20,21 *Fullonica di Veranius* *Hypsaeus*

NO ACCESS. Building work

Latrine in kitchen in 20 (Eschebach p186)

VI.8.3,4,5,6 *Casa del Poetica tragico*

LATRINE in the kitchen area of the house in a narrow room with masonry marks indicating the seat level for a toilet

VI.8.7, 8 *'Large thermopolium'*

LATRINE in the northernmost part of this property in a room divided into two, the western part of which contains a plaster niche.

(Eschebach p183:Fiorelli p61)

There is a DOWN PIPE of diameter 25cm.In the room to the east of the latrine This probably is from an upper storey latrine and it may drain into a common cesspit.

VI.8.9,10 *'Refreshment room'*

DOWN PIPE of diameter 20cm, in the western wall.

Latrine in 9 (Eschebach p184)

VI.8.11 *'Bottega'*

In Room 2 there is an arch through which a DOWN PIPE is accessing a pit.

VI.8.12,13 *'Thermopolium'*

LATRINE with a plaster niche.

(Fiorelli 1875:61)

VI.8.14 *'Sacrarium'*

Nothing was found in this property (Fiorelli p61).

VI.8.15 *'Shop'*

Nothing.

VI.8.16 *'Shop'*

Nothing.

VI.8.17 *'Shop'*

Nothing.

VI.8.18 *'Shop'*

Nothing.

VI.8.19 *Shop*

DOWN PIPE on the southern wall of this small single roomed property

VI.8.23,24 *Casa della Fontana Piccola*

LATRINE in this house in room 20. It has been described as being in the kitchen 'next to the hearth with a very flimsy wall dividing it from the rest of the room' There is a window adjacent (Fröhlich 1996)

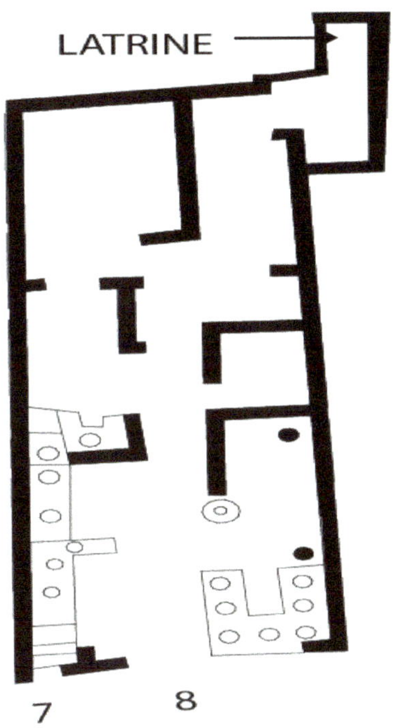

VI.8.7,8 Latrine room

VI.8.7,8 Down pipe: Room 22

VI.8.8 Down pipe: Room 18 'A'

VI.8.8 Down pipe Room 18'B'

VI.8.9,10 Down pipe

VI.8.11 Arch & Down Pipe Room 2

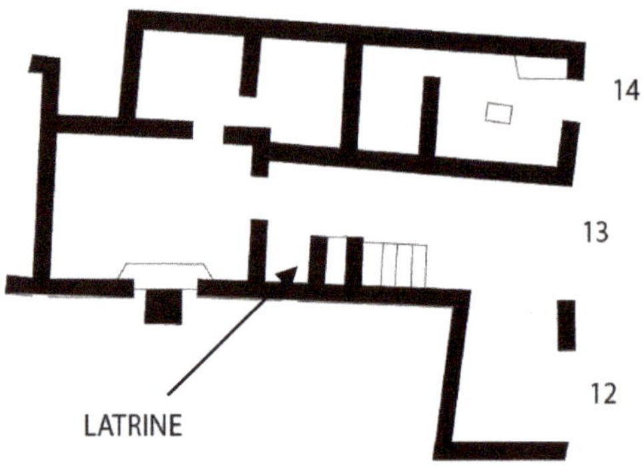

Latrine in VI.8.12,13

VI.8.17

VI.8.19 Down pipe

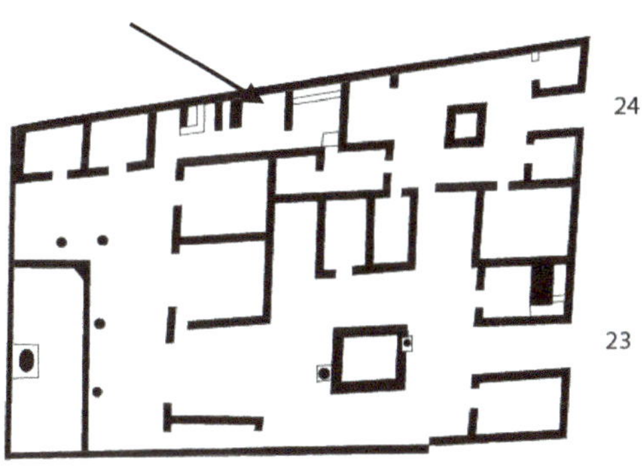

VI.8.23,24

<u>REGIO VI. 9</u>

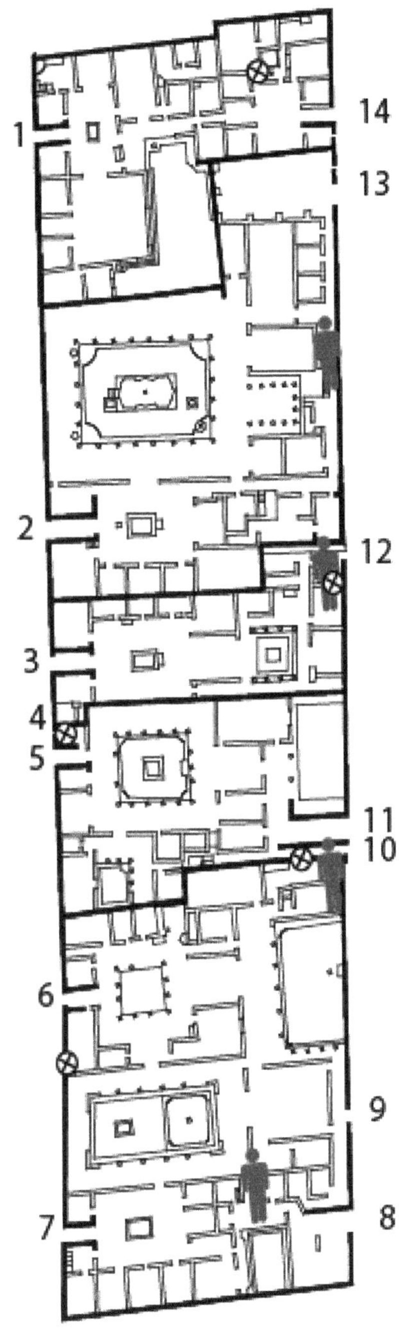

There are six major properties in this *insula*. Only four down pipes have been observed, two of which may be for water.

VI.9.1,14 ***Casa d'Iside ed Io***

Nothing obvious.

(Eschebach latrine in 1: Packer Room 4 kitchen in 1)

VI.9.2,13 *The Casa di Meleagro*

To the rear of this property are a number of rooms by one of which is a deep pit probably the cesspit for the latrine.

Latrine in 2 (Eschebach p189: Fiorelli p64)

VI.9.3-5,10-12 ***Casa del Centauro***

LATRINE with DOWN PIPE in room 15 of this property. (Eschebach latrine).

Slots, for the seat are cut into the masonry of the wall and of the cooking surface.

VI.9.6,9 ***Casa dei Dioscuri***

LATRINE with masonry supports over a metre apart. (Eschebach p191: Fiorelli p66)

Thèdenat mentions the two windows. Jansen says three

VI.9.7,8 ***Casa Cn. Caetroni Eutychi***

NO ACCESS Building work in progress. (Eschebach p 191:Fiorelli p66)

VI.9.1,14 Down pipe: Room 29

VI.9.2,13

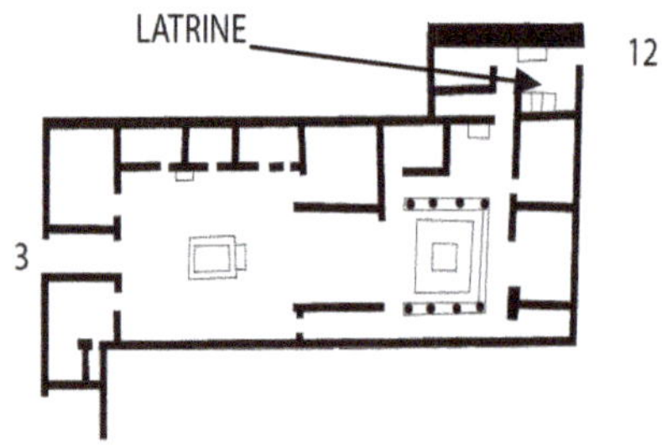

VI.9.3,12

VI.9.3,12 Down pipe visible in outside of arch of eastern wall of latrine

• VI.9.4 Room 15

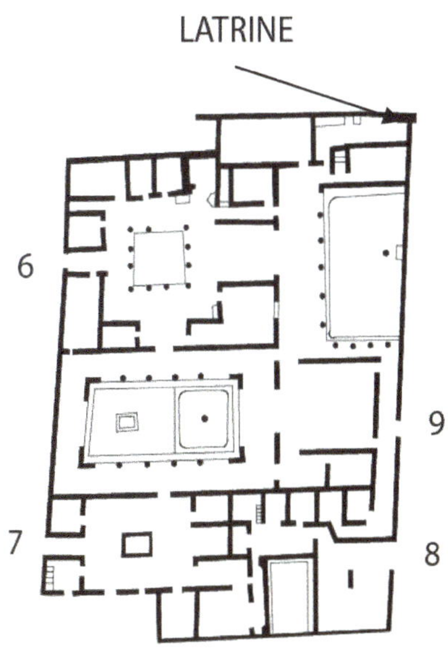

VI.9.6

VI.9.6 Room F

REGIO VI.10

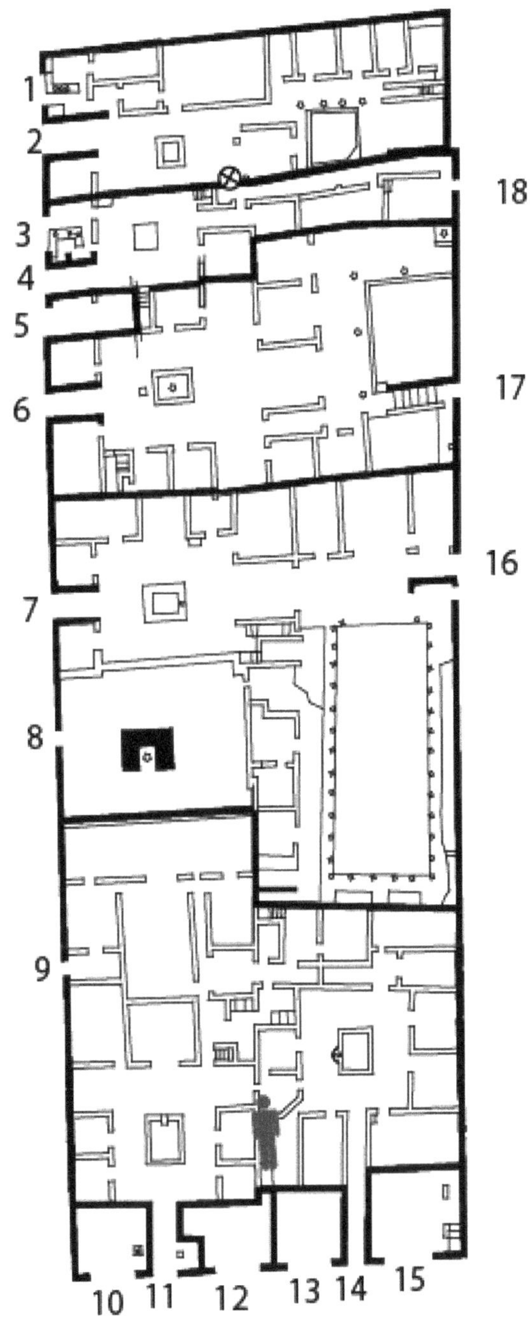

There are six houses and five shops in this *insula*. Only one down pipe and one latrine have been identified.

VI.10.1,2 ***'Caupona'***

Nothing.

Latrine in 2 (Eschebach p192)

VI.10.3,4,18 ***'Thermopolium'***

Nothing, but Fiorelli says that there is a latrine by entrance #18 (Fiorelli **p**67). (Eschebach p193)

VI.10.5 ***'Shop'***

Nothing.

VI.10.6,17 ***Casa di Pomponius***

NO ACCESS

VI.10.7,8,16 ***Casa di Nettuno***

This is a sizeable house but there is no obvious latrine.

Latrine in 7 (Eschebach p194)

VI.10.9,11 ***Casa del Naviglio***

A very badly damaged property. No obvious latrine (Eschebach latrines in 9 p194 & 11 p195:Fiorelli latrines in 9 & 11 p68).

VI.10.10 ***'Shop'***

Nothing.

VI.10.12 ***'Shop'***

Nothing.

VI.10.13 ***'Shop'***

There is a scar on the eastern wall possibly for a down pipe. Otherwise nothing.

VI.10.14 ***'House'***

LATRINE in the kitchen area is the latrine which was excavated by Sogliano (Sogliano 1900).

VI.10.15 ***'Large shop'***

Nothing

VI.10.1Water pipe in south wall room 4

VI.10.14 Latrine

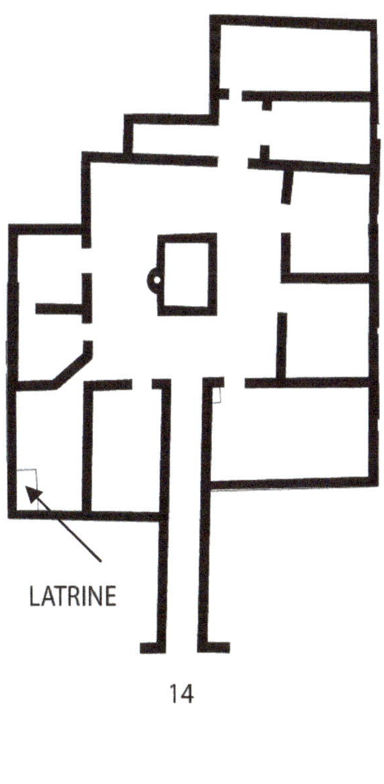

REGIO VI.11

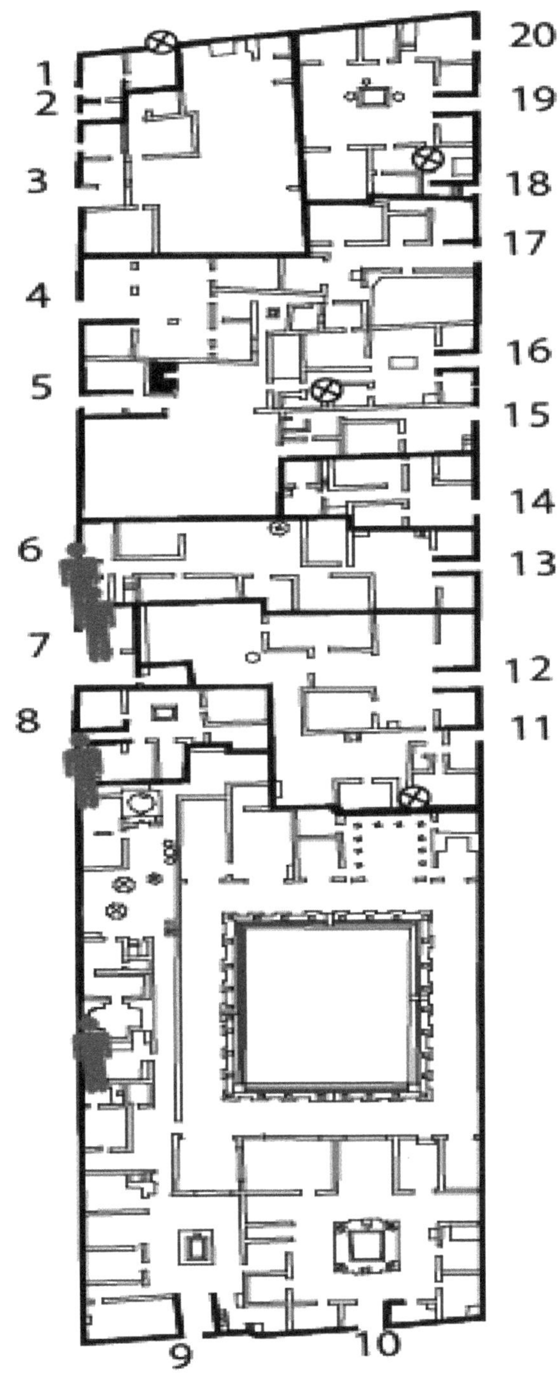

There are ten properties in this *insula,* almost half the area of which is occupied by the Casa di Labirinto. The northern section is heavily overgrown making it impossible to find the latrines. Four down pipes were visible in this *insula*

VI.11.1,2 ***'Shop'***

Nothing

VI.11.3. ***'Shop with garden'***

Nothing.

VI.11.4,5,15,16,17 ***'Workshop'***

Nothing

Latrine in 15 (Eschebach p198)

VI.11.6,13 ***Officina vestiaria***

LATRINE in the small room to the right of the entrance #6. (Eschebach p198: Fiorelli p70)

VI.11.7 ***'Small house'***

NO ACCESS due to building work. Fiorelli mentions a latrine (Eschebach p198: Fiorelli p69).

VI.11.8. ***House of Eutychus***

LATRINE This latrine appears to be the one for the bakery and perhaps for the bathhouse of the *Casa di Labirinto* (see below)

VI.11.9,10 The ***Casa di Labirinto***

A very large house, currently undergoing repairs. There is a possible site of a latrine in the kitchen area but it would appear, since it is accessible from the main part of the house, that the latrine is in VI.11.8. This latrine, in room 56, is the one in the Hauser series (Strocka 1991, 56). It has one pedestal and one slot.

There is a window and probably a door

VI.11.11,12 ***'Reienhaus'***

Nothing.

Latrine (Eschebach p200)

VI.11.14 ***'Small house with workshop'***

Nothing.

VI.11.18,19,20 ***'Hospitium'***

This area is totally overgrown

VI. 11. 1 Down pipe: Outer north wall

VI.11.6,13 Latrine, room 3

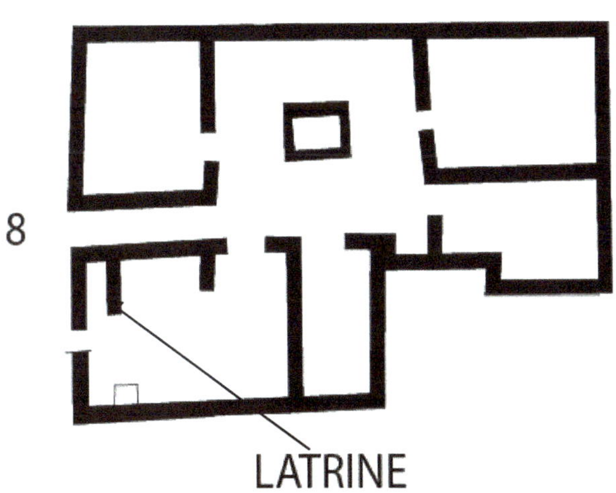

VI.11.8

VI. 11. 15 Down pipe: Room 11, east wall

VI. 11. 18 Down pipe Room 11 east wall

REGIO VI.12

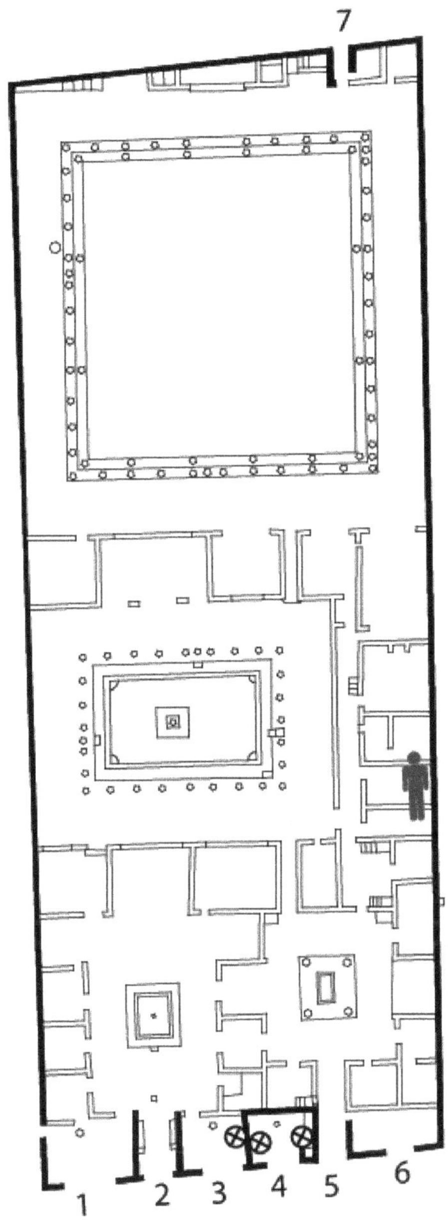

This *insula* is occupied almost totally by the Casa del Fauno. There are a few subdivisions of the property where it borders on the street to the south, and here there are three down pipes indicating upper storeys.

VI.12.1,2,3,5,6,7 ***Casa del Fauno***

LATRINE in the service area to the east of the largest peristyle. There is only one latrine for the whole house. The masonry supports are 1m 36cm apart suggesting that it may have been a two (or even three) seat toilet. It has been stated that the waste water of the *caldarium* to the north was used to flush this latrine (de Haan 2001:44) though there is no obvious evidence of this. Contrarily Jansen says that piped water was brought in from outside through the wall (Jansen 2001:40). A hole above the southern masonry seat support can be seen in the illustration. (Fiorelli p71)

VI.12.3

DOWN PIPE

VI.12.4 ***'Shop'***

No evidence of a latrine.

TWO DOWN PIPES

VI.12.2 Casa del Fauno

VI.12.3 Down pipe

VI.12.4 Down pipe

REGIO VI.13

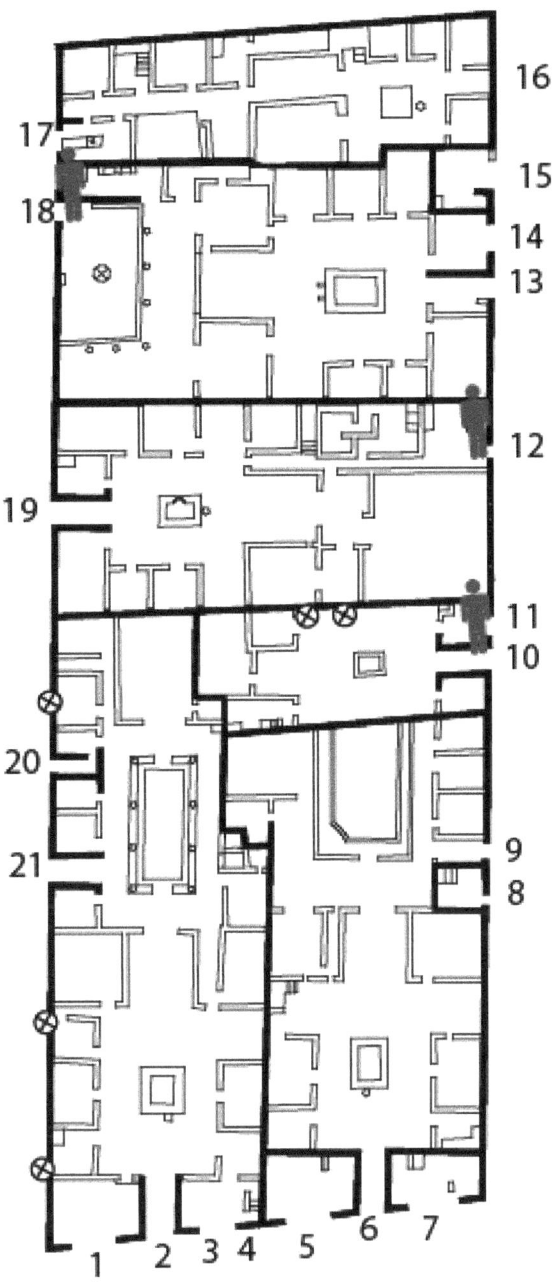

This *insula* has six main properties and some shops.

VI.13.1,2,3,4,20,21 ***Casa del Gruppo dei Vasi di Vetro***

Thèdenat (1910:96) mentions the latrine in this house as having one window. However there was no evidence in this house of a latrine.

Latrine in 2 (Eschebach p203)

DOWN PIPE in outer wall by #20

VI.13.5 ***'Shop'***

Nothing.

VI.13.6,8,9 ***Casa del Forno***

A large house but nothing obvious in the way of a latrine.

VI.13.7 ***'Shop'***

Nothing.

VI.13.10,11 ***'House with restaurant'***

LATRINE with a sloping floor, in a small room just inside the doorway to the north

(Eschebach p205)

TWO DOWN PIPES on north wall in room b

VI.13.12,19 ***Casa di Sextus Pompeius Axiochus***

LATRINE in the small room to the north of the entrance #12.

VI.13.13,18 ***'Dwelling house'***

LATRINE in a room at the north-western part of the house. The slot for the seat is clearly visible in the wall to the west

(Eschebach p206)

VI.13.14 ***'Shop'***

Nothing.

VI.13.15 ***'Shop'***

Nothing.

VI.13.16,17 ***Casa di Gavius Proculus***

Latrine in 16 (Eschebach p207)

VI.13.10,11

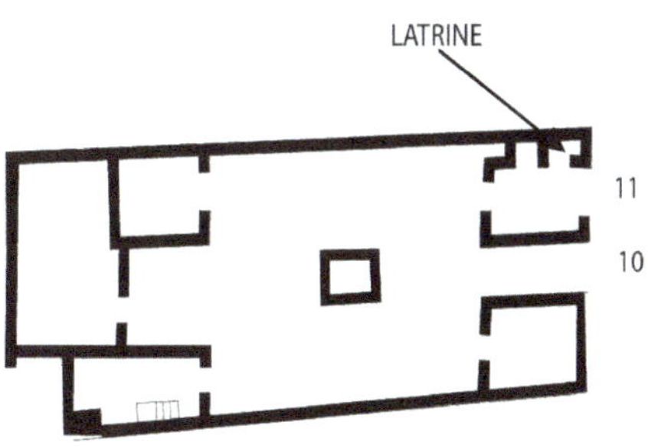

VI.13.10,11 Two down pipes: north wall, room b

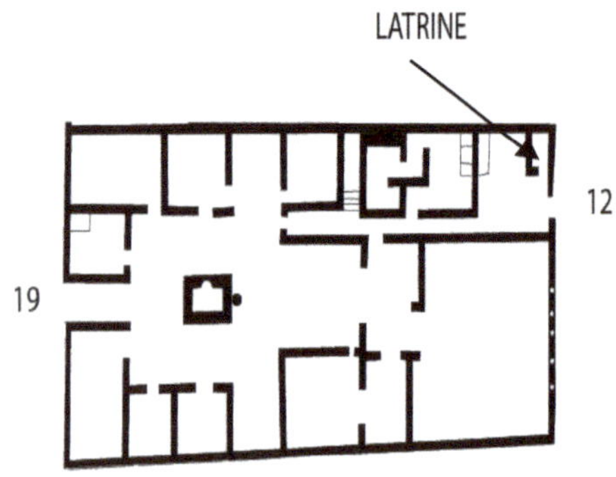

VI.13.12,19

VI.13.13,18 Latrine

VI.13.20 Down pipe

VI.13.21Downpipe

REGIO VI.14

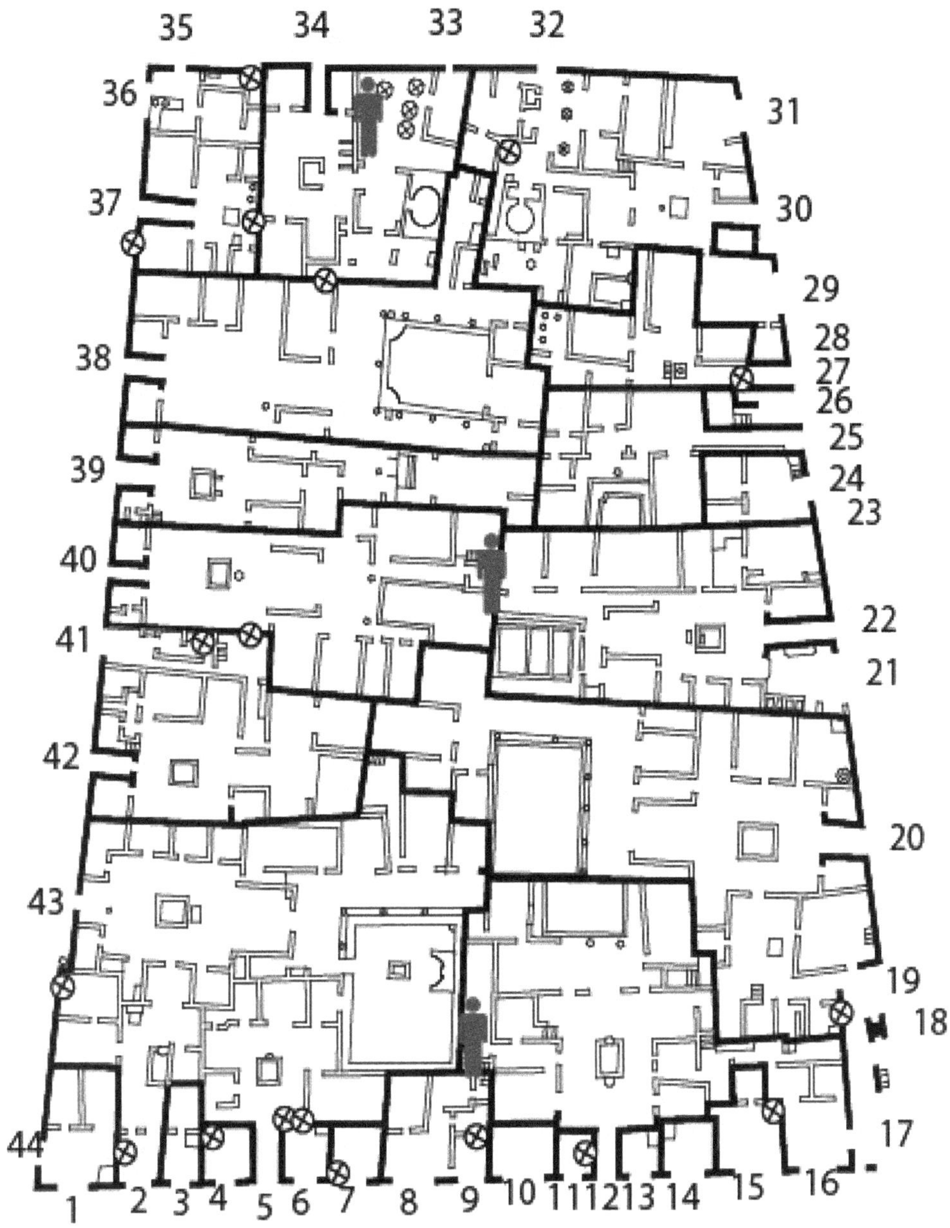

This *insula* has thirteen sizeable properties and a considerable number of smaller ones including seven single-room shops on the southern street aspect. There is a conspicuous lack of latrines within this *insula* but there are seventeen down pipes a considerable number of which were probably serving upper storey latrines.

VI.14.1 *'Popina'*

Nothing.

VI.14.2 *'Fish shop'*

DOWN PIPE of 23cm external diameter in the south western corner.

No evidence of a latrine in this small property

VI.14.3 *'Shop'*

Nothing.

VI.14.4 *'Shop'*

DOWN PIPE of 20cm external diameter in the north wall.

VI.14.5,43 *Casa del Marte*

Nothing.

VI.14.6 *'Shop'*

Nothing.

VI.14.7 *'Shop'*

DOWN PIPE In south east corner.

VI.14.8,9 *'Business shop'*

DOWN PIPE in the eastern wall.

Latrine near entrance to 9 (Eschebach p209)

VI.14.10 *'Shop'*

Nothing.

VI.14.11 *'Shop'*

PIPE cut at floor level .

VI.14.12 *Casa del Numisius Rarus*

LATRINE in a small room in the secluded south east part of the house, demarcated by some plaster marks. (Eschebach p210)

VI.14.13 *'Shop'*

Nothing.

VI.14.14 *'Shop'*

Nothing.

VI.14.15 *'Shop'*

DOWN PIPE in NE corner room 1

? Latrine (Eschebach p211)

VI.14.16 *'Shop'*

Nothing.

VI.14.17,18-20 *Forum Gallinarium*

Nothing.

VI.14.21,22 *Fullonica del M. Vesonius Primus*

Nothing.

Latrine in 22 (Eschebach p213)

VI.14.23,24 *'Shop'*

Nothing.

Latrine in 24 (Eschebach p213)

VI.14.25 *'Dyehouse'*

Nothing.

Latrine (Eschebach p213)

VI.14.26 *'Shop'*

Nothing.

Latrine (Eschebach p214)

VI.14.27 *Casa del m. Memmius Auctus*

NO ACCESS

DOWNPIPE in the southern wall, photographed from the barred and locked door.

VI.14.28-32 *Casa di Laocoonte*

DOWN PIPE placed within an inner corner of the north west of the property.

VI.14.29 *'Shop'*

Nothing.

VI.14.33,34 *'Bakery'*

There is a small feature by the western wall of property #33 which might be a latrine.
(Eschebach p215)

VI.14.35,36 *Caupona del Salvius*
DOWN PIPE in room c.
Latrine in 36 (Eschebach p215: Packer latrine in kitchen p45)

VI.14.37 *'Workshop'*
Water pipe in east wall.

VI.14.38 *Casa del C. Poppaeus* *Firmus*
Nothing.

VI.14.39 *Casa Lucrum Gaudium*
Nothing.

VI.14.40 *'Workshop'*
Latrine in kitchen (Eschebach p217)
LATRINE in this small room under the staircase in this property says Thèdenat. Possibly under the rubble here.

VI.14.41,42 *Casa del Inventus*
DOWN PIPE set in the north wall of this property.
The latrine may have been in the small narrow room on the west side of the house.

VI.14.43 *House of the brothel*

VI.14.2 Down pipe

VI.14.4 Downpipe

VI.14.7 Down pipe

VI.14.11 Pipe

VI.14.12

VI.14.15 Pipe

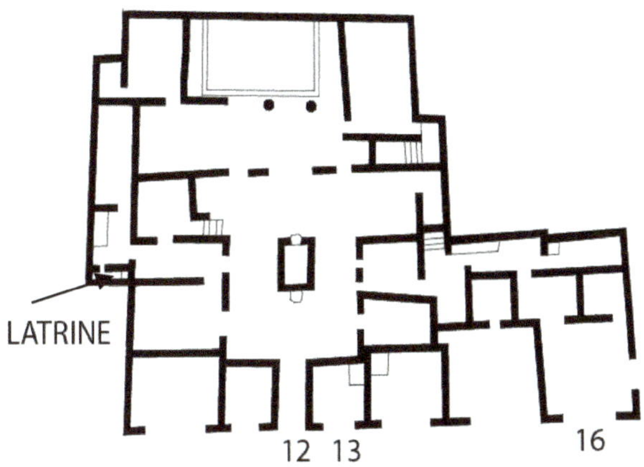

VI.14.17 Down pipe in outer wall

VI.14.27 Down pipe

VI.14.30,31,32

VI.14.33,34

VI.14.35,36

VI.14.40 Under staircase

VI.14.41 Room 4

VI.14.42

REGIO VI.15

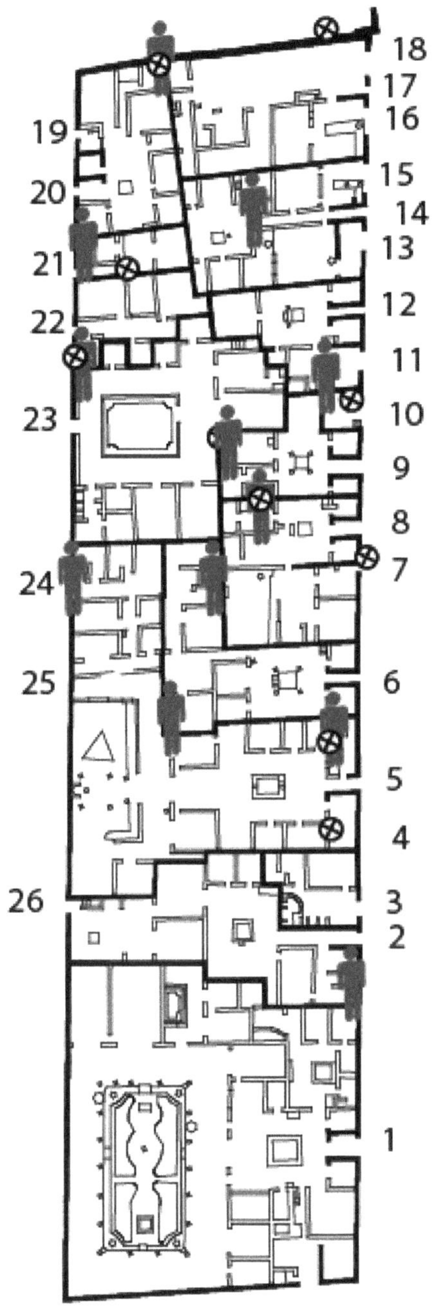

Insula fifteen of Regio VI is dominated at its southern end by the Casa di Vettii. It also has a number of houses including the Casa di Principe Napoli with some beautiful decoration. Some of the houses are quite small but well appointed. There are two upper storey latrines

VI.15.1,27 ***Casa di Vettii***

Latrine by door #27in room 2 (Eschebach p218)

Allegedly there is a latrine associated with the kitchen. Building work has precluded access

VI.15.2,26 ***Casa d' Appuleia***

Possible in the second room to the south of entrance #2 (Eschebach p219)

VI.15.3 ***'Fullonica'***

NO ACCESS.

VI.15.4,5,24,25 *Casa M. Pupius Rufus*

These two properties are connected (not clearly shown in the plan) There are two latrines

To the north of entrance #5 there is a complex of rooms including the kitchen, north of which is the small room for the LATRINE. Thédenat comments that the latrine empties onto the road.

LATRINE in a room with slots in the walls for a latrine seat, to the north of blocked entrance #24 (Eschebach p219)

VI.15.6 ***Casa del Caesius Valens***

LATRINE (Eschebach p220)

VI.15.7,8 ***Casa del Principe Napoli***

LATRINE in a small dark kitchen area, room g. There is a down pipe behind the latrine (Eschebach p220)

Since the publication of the description of the house by Strocka (1984,21) the southern masonry support for the seat has collapsed.

VI.15.9 House

LATRINE (Eschebach p221)

VI.15.10 ***'Shop'***

DOWN PIPE within the north east corner measuring 19cm external diameter.

VI.15.11,12 ***'House with accommodation'***

Possible in a small room in the southern part of the property. (Eschebach p221)

VI.15.13,14,15 ***Casa della Matrona ignota***

LATRINE in a small room in the interior of the house. The plaster detail of the side walls, and the slot for the seat confirm that this was the latrine. (Eschebach p222)

VI.15.16,17,18 ***'Popina'***

Possible in a long narrow room at the rear of this property.

VI.15.19,20 ***Casa del M. Stlaborius Auctus***

UPPER STOREY LATRINE

Latrine (Eschebach p223)

VI.15.21 ***'Workshop'***

LATRINE in a room to the north of the entrance with a niche in the western wall

VI.15.22 ***House of Cinnius Fortunatus***

UPPER STOREY LATRINE

VI.15.23 *'Hospitium'*

Possible in the large kitchen area in the south western part of the property which was filled with rubble. (Eschebach p223)

VI.15.24 Masonry blocked doorway.

VI.15.25 Masonry blocked doorway.

VI.15.2,26

VI.15.4

VI.15.5 Latrine with downpipe in masonry

VI.15.5 (24)

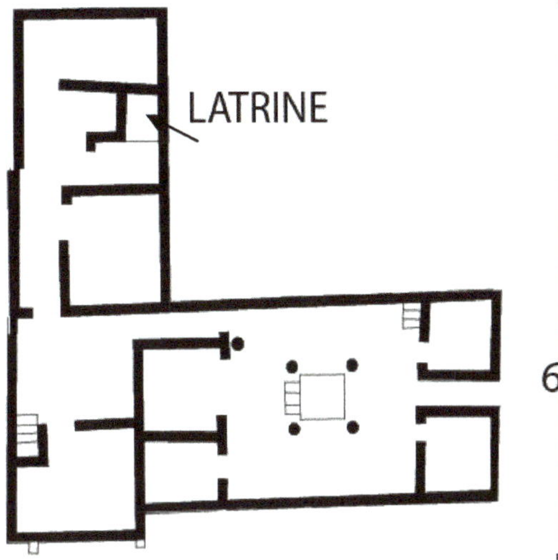

VI.15.6

VI.15.7 Down pipe at base of wall

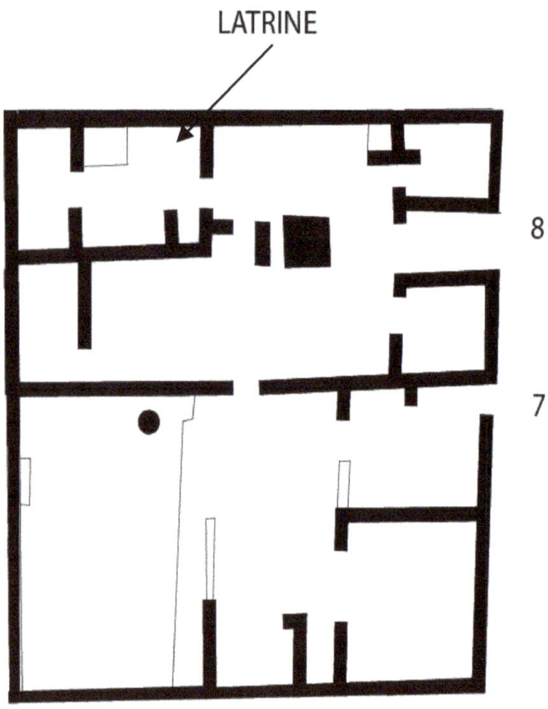

VI.15.7,8

VI.15.8 Latrine with sloping, narrow bore pipe

VI.15.10 Down pipe

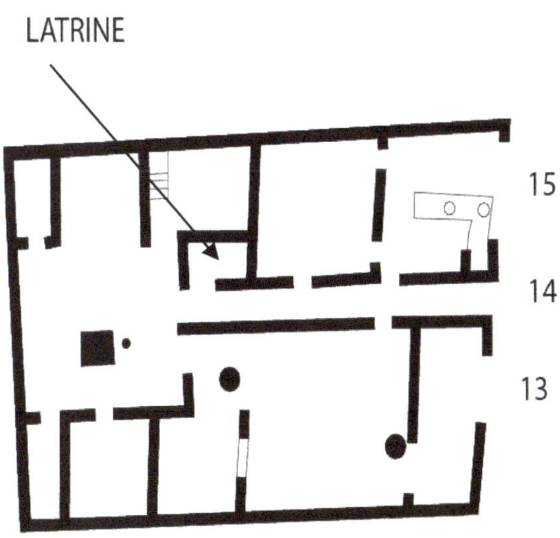

VI.15.13,14,15

VI.15 18 Pipe in outer wall

VI.15.16,17,18

VI.15.19

VI.15.20

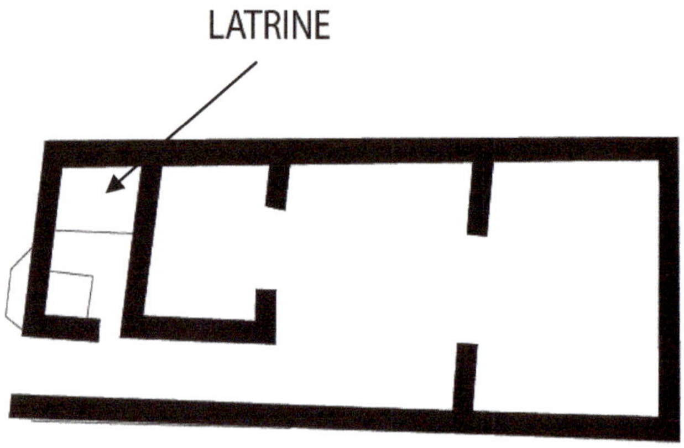

VI.15.21

VI.15.22 Latrine with down pipe

VI.15.22 Upper storey latrine

REGIO VI. 16

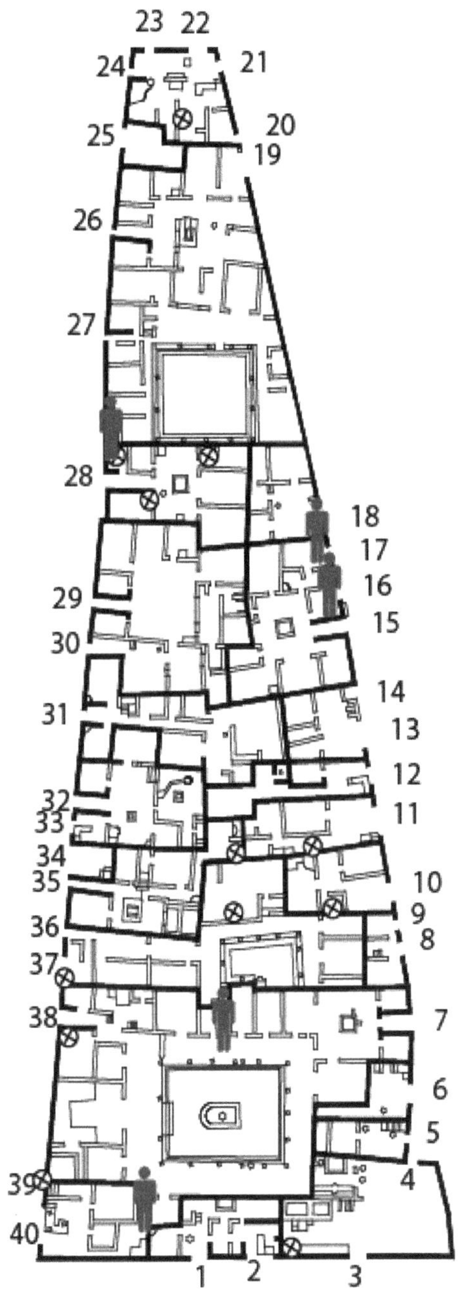

Insula 16 is triangular in shape with the base of the triangle to the south. It is dominated by two large properties, the Casa degli Amorini Dorati at the base and the Casa dell' Ara Massima towards the north. Each of these has been studied previously and the latrines recorded (Seiler 1992 Stemmer 1992).
There are five small premises bordering the street and the south eastern corner. The remaining properties are small houses.

Relatively few latrines for the number of properties but some evidence of upper storey latrines

VI.16.1,2 *'Thermopolium'*

Nothing.

VI.16.3,4,5 *'Fullonica'*

Nothing. The water supply to the fullonica runs from the Casa degli Amorini Dorati

VI.16.6 *'Fullonica'*

Nothing.

VI.16.7,38 *Casa degli Amorini Dorati*

Latrine in kitchen in 7(Eschebach p226)

LATRINE in a narrow room K, in the north of the peristyles. One pedestal (Seiler 1992.52)

Latrine in Taverna, entrance 38, room x (Seiler.1992.68) Not visible.

DOWN PIPE

VI.16.8,9 *'Shop'*

Nothing.

VI.16.10 *'Small house'*

Latrine (Eschebach p226)

Water pipe on south wall of room c.

I.16.11.35 *'Workshop'*

Latrine in kitchen (Eschebach p227)

Nothing.

VI.16.12 *'Thermopolium'*

Nothing.

Latrine in kitchen (Eschebach p227)

VI.16.13,14 *'Shop'*

Nothing.

VI.16.15.16.17 *Casa dell' Ara Massima*

LATRINE in the kitchen area to the left of entrance #17 which has been newly restored. Room L (Stemmer 1992,35) (Eschebach p227)

VI.16.18 *'Workshop'*

Latrine (Eschebach p228)

The masonry support on the south wall suggests that this was the latrine.

VI.16.19,26,27 *'Business house'*

LATRINE A narrow room in the south west of the property with a mark on the western wall suggesting the position of the seat.

(Eschebach p229)

VI.16.20,21,22,23,24 *'Thermopolium'*

Latrines in 21 and 24 (Eschebach p228)

DOWN PIPE in wall of room b

VI.16.25 *'Shop'*

Nothing.

VI.16.28 *Casa dalla Caccia di Tori*

There are three down pipes in this property. Two of these may be associated with upper storey latrines.

DOWN PIPE A room B

External diameter 18cm

DOWN PIPE B room G

External diameter 18cm

DOWN PIPE going into cistern

External diameter 10cm.

VI.16.29,30 *'Workshop'*

Nothing

Latrine in 29 (Eschebach p230)

VI.16.31 *'Workshop'*

Nothing.

VI.16.32,33 *'Popina'*

Nothing.

VI.16.34 *'Shop'*

Nothing.

VI.16.35 *'Small house'*

Nothing.

VI.16.36,37 *Casa del Q. Poppaeus* Sabinus

Possible in the kitchen area of this property

VI.16.39.40 *Thermopolium and Popina of Felix and Dorus*

LATRINE, room G with DOWN PIPE in floor

Latrine in 40 (Eschebach p232)

VI.16.10 Waterpipe: Room c

VI.16.11 Pipe in south wall room C

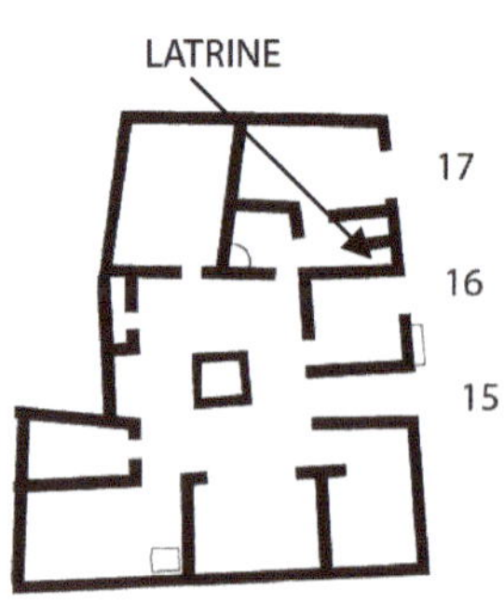

VI.16.17

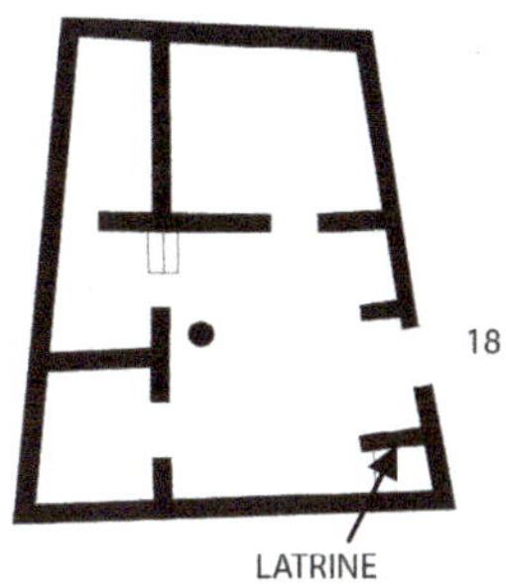

VI.16.18

V.16.21 Down pipe: Room b

VI.16.19, 25, 26,27 Latrine in room Z

VI.16.28 Down pipe B: Room G

VI.16.28 Down pipe A: Room B

VI.16.28 Pipe going into cistern

VI.16.35 Downpipe: Room G

VI.16.36,37

VI.16.37 Pipe in brick door pillar

VI.16.38 Down pipe: Room T

VI.16 Rain water pipe in wall to north of #39

VI.16.7, 38 (wrong address on board)

VI.16.40 Latrine with down pipe in floor: Room G

REGIO VII (Insulae 1-15)

REGIO VII (Insulae 1-15)

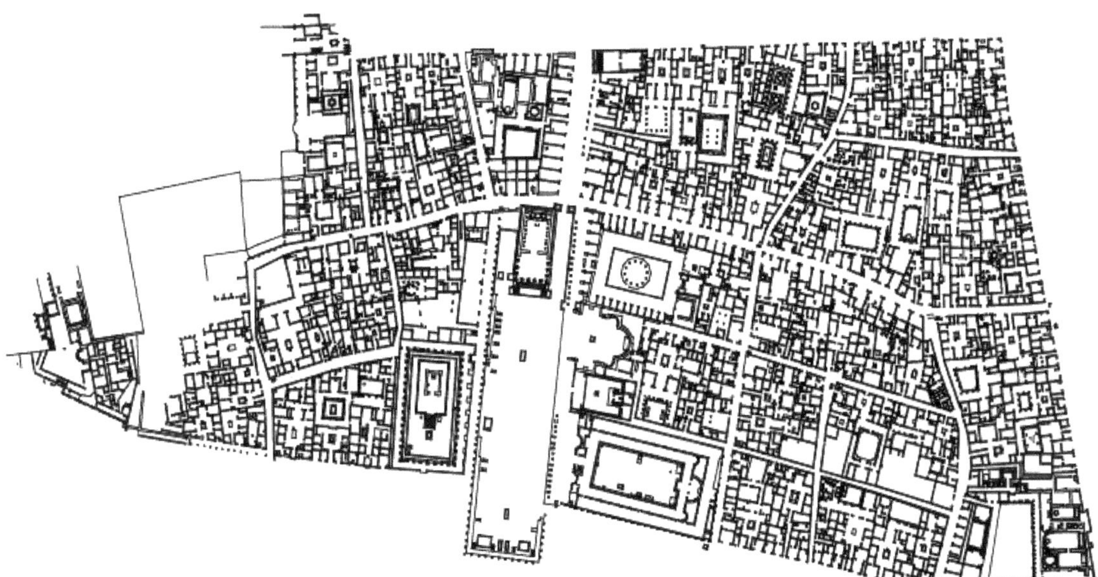

REGIO VII.1

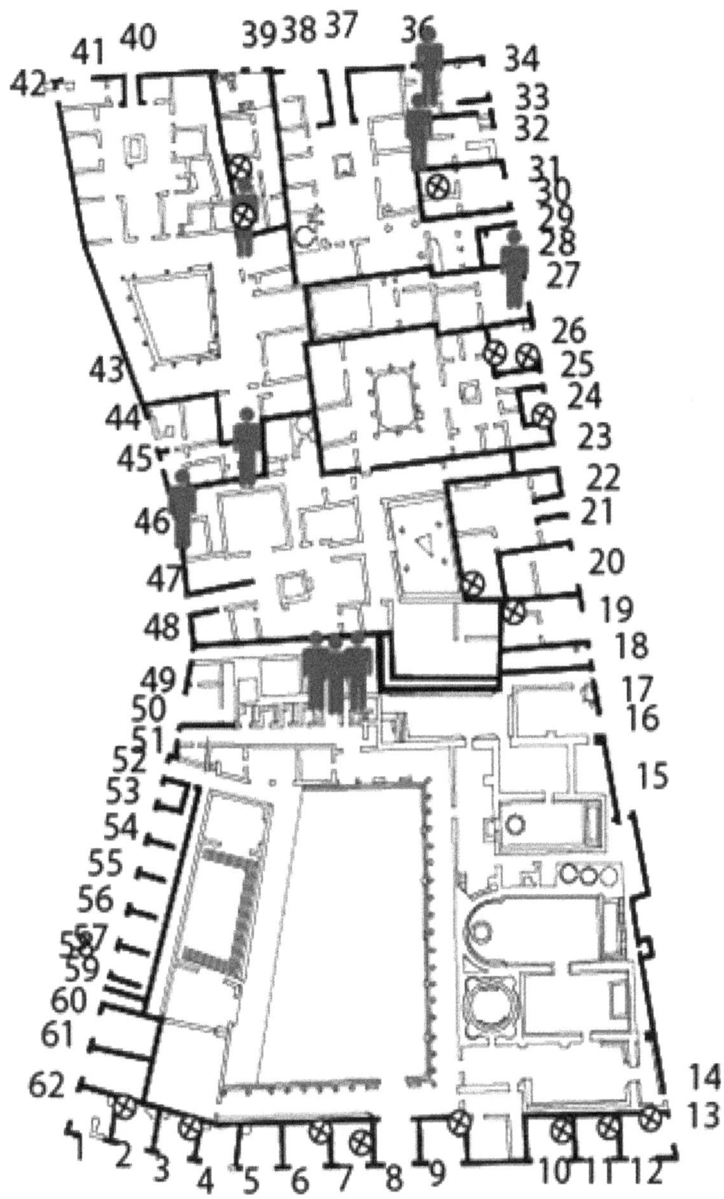

The Stabian baths occupies almost half of this *insula*. To the east and the south of the baths are shops, the southern group having much evidence of second storeys containing latrines. Many of the shops have down pipes running in the north walls.

To the north of the *thermae* there are four large properties and thirteen smaller ones. Eschebach notes a number of latrines in this *insula* the evidence for which is lacking without cleaning/excavation

VII. 1. 1,62 *Caupona*

DOWN PIPE. Also visible in VII.1.61

VII. 1. 2 *Shop of Proculus*

Nothing

VII. 1. 3 *Shop of Clodius Sagarius*

Latrine (Eschebach p241)

DOWN PIPE in northeast corner

VII. 1. 4 *Shop of Sagata*

Nothing

VII. 1. 5 *Shop of Sestius Venustius*

Nothing

VII. 1. 6 *Shop of Stronnius Valens*

DOWN PIPE and possible Latrine

VII.1.7 *Shop of M Ampullius Cosmus*

Eschebach latrine

DOWNPIPE in northeast corner

VII. 1. 8, 14, 15-17, 48, 50, 51 Stabian Baths

MULTI-SEATER LATRINE (Eschebach p241)

(NO PHOTOGRAPH. BUILDERS ON SITE)

VII. 1. 9 *Shop of Sestius Proculus*

DOWN PIPE in northeast corner

VII. 1. 10 *Shop*

DOWN PIPE in northeast corner

VII. 1. 11 *Shop of Sabinus*

DOWN PIPE in northeast corner

VII. 1. 12, 13 *Eckladen*

DOWN PIPE in northeast corner

VII. 1. 14 Entry to Stabian Baths

VII. 1. 15 Entry to Stabian Baths

VII. 1. 16 *Workshop*

Nothing

VII. 1. 17 Entry to Stabian Baths

VII. 1. 18 *Small room*

Nothing

VII. 1. 19 *Shop*

DOWN PIPE and POSSIBLE LATRINE

VII. 1. 20 *Shop*

Nothing

VII. 1. 21, 22 *Craft workshop*

(Eschebach p246:Fiorelli p75)

DOWN PIPE in southwest corner of 22

VII. 1. 23

Public latrine (Eschebach p246)

Nothing

VII. 1, 24 *Shop*

DOWN PIPE in outer wall above arch.

VII. 1. 25, 46, 47 *House of PP Vedii Siricus and Numinianus (House of the Prince of Russia)*

LATRINE.

Latrine in 46 (Eschebach p246)

? Eschebach latrine in 47 (Fiorelli p79)

VII. 1. 26 *Large shop of Ceratus*

TWO DOWN PIPES

VII. 1. 27 *Business shop of Vibia and Ameia* (Eschebach p248 two latrines: Fiorelli p76 two latrines)

LATRINE to right of entrance

VII. 1. 28, 29 *Shop*

Nothing

VII. 1. 30 *Stable*

Nothing

VII. 1. 31 *Smithy*

DOWN PIPE in room 2

Latrine (Eschebach p248)

VII. 1. 32 *Bar*

LATRINE in room 3

(Eschebach p249:Fiorelli p76)

VII. 1. 33-35 *Business shop of Clodius Nymphodotus*

Latrine in 35 (Eschebach p249)

Nothing found

VII. 1. 36, 37 *Pistrinum*

36 LATRINE (Eschebach p249)

VII. 1. 38, 39 *Bar*

39 LATRINE (Eschebach p250)

VII. 1. 40-43 *House of M. Caesius Blandus*

Nothing

VII. 1. 44, 45, 45a *Inn of Sittii*

LATRINE in 45a (Eschebach p251)

VII. 1. 49 *Shop of M Stabius Chryseros*

Nothing

VII. 1. 53-58 *Probable shop associated with the Baths*

Nothing

VII. 1. 59 Stair entrance with possible public latrine (Eschebach p253)

VII. 1. 60 *Large shop*

Nothing

VII. 1. 1,61 *Large shop*

Nothing

VII. 1. 3 Down pipe

VII. 1. 6 Down pipe

VII. 1. 7 Down pipe

VII.1.9 Downpipe in corner of stairs

VII. 1. 10 Down pipe

VII. 1. 11 Down pipe

VII. 1. 12 Downpipe

VII. 1. 19

VII. 1. 21, 22

VII.1.24 Downpipe

VII.1.26 Downpipe

VII.1.26 Downpipe B

VII.1.27 Latrine

VII.1.31 Downpipe

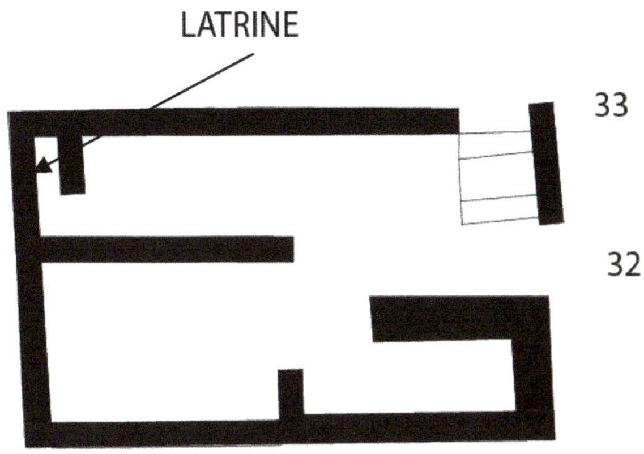

VII. 1. 32

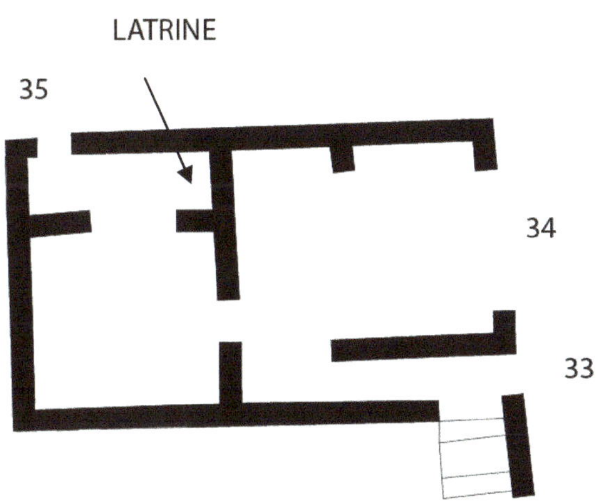

VII. 1. 36

VII. 1. 38,39
Above LATRINE
Left DOWNPIPE

VII.1.38 Downpipe B

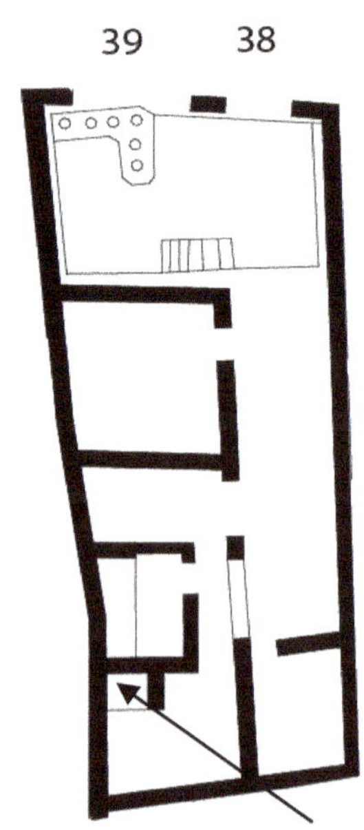

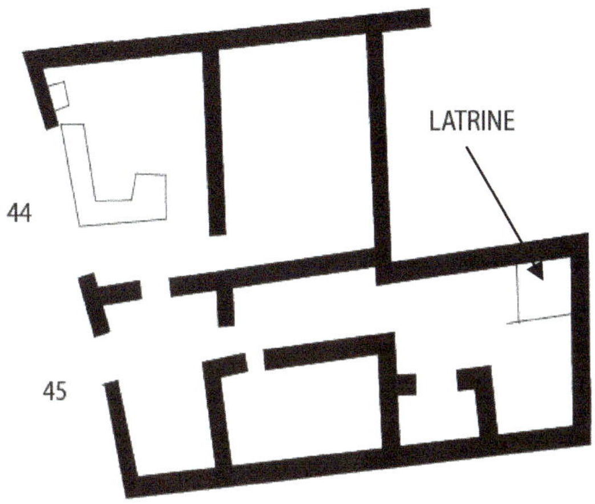

VII. 1. 44, 45 room d

VII.1.46 Latrine

VII. 1. 62 Down pipe

REGIO VII.2

This insula consists of thirty two properties. The down pipes in the southern properties suggest upper storey accommodation. There is a most interesting upper storey latrine in VII.2.18

VII. 2. 1 ***Shop***

Nothing

VII. 2. 2 ***Shop of Marcellus***

Nothing

VII. 2. 3, 6, 7 ***House of T.T.Tarentii***

LATRINE in room e (Eschebach p254)

VII. 2. 4 ***Shop***

Nothing

VII. 2. 5 ***Shop of Sabinus***

Nothing

VII. 2. 7 ***Large shop***

Nothing

VII. 2. 8, 9 ***Stairs and access***

LATRINE (Eschebach p255: Fiorelli p80))

VII. 2. 10 ***Shop***

POSSIBLE

VII. 2. 11 ***'Infectorium' of Ubonius***

Nothing

VII. 2. 12 ***Bar***

DOWN PIPE N.E corner

VII. 2. 13 ***Shop***

Nothing

VII. 2. 14, 15 ***House of Optatio***

Nothing

Latrine (Eschebach p256)

VII. 2. 16 ***House of Gaius Rufus (House of the seven skeletons)***

LATRINE

VII. 2. 17 ***Small shop***

Nothing

VII. 2. 18, 19, 42 ***House of CC Vibi and Italus***

UPPER STOREY LATRINE

LATRINE in room v (Eschebach p258)

VII. 2. 20, 21, 41 ***House of NN Popidii, Priscus, Ampliatus***

UPPER STOREY LATRINE with down pipe

LATRINE in room b (Eschebach p258)

VII. 2. 22 ***'Eckhaus' of M. Fabius Lalus***

Nothing

Latrine (Eschebach p259: Fiorelli p83)

VII. 2. 23 ***Casa di Amore punito***

DOWNPIPE in room g south wall

Possible latrine (Eschebach p259)

VII. 2. 24-26 ***Casa della Caccia Nuova***

Nothing

Latrine under stairs in 24 (Eschebach p260: Fiorelli p83)

VII. 2. 27, 29-31, 35 ***Casa di Mercurio***

LATRINE in room g

Public latrine in 28 (Eschebach p260: Fiorelli p83)

VII. 2. 32, 33 ***Taberna and caupona of Philippus***

Nothing

VII. 2. 34 ***Shop of Philippus***

Nothing

VII. 2. 37-39 ***Casa delle Quadraghe***

LATRINE off room b (Eschebach p262: Fiorelli p84)

VII. 2. 40 ***Large shop***

Nothing

VII. 2. 42 ***See House of Vibi and Italus***

DOWNPIPE in room 1

VII. 2. 43 ***Shop of Magonius***

Nothing

VII. 2. 44, 45 ***Casa dell'Orso***

LATRINE in room g (Ehrhardt 1988,36)

(Eschebach p263)

VII. 2. 46 ***Large shop***

DOWNPIPE

VII. 2. 47 ***'PUBLIC LATRINE'***

(Eschebach p263:Fiorelli p85)

Requires 'cleaning'

VII. 2. 48, 49 ***Workshop of D. Caprasius Primus***

Nothing

(Eschebach p263)

VII. 2. 50 ***Large shop***

Nothing

VII. 2. 51 ***House of the Suetti***

DOWNPIPE

VII. 2. 52, 53 ***Small workshop***

Nothing

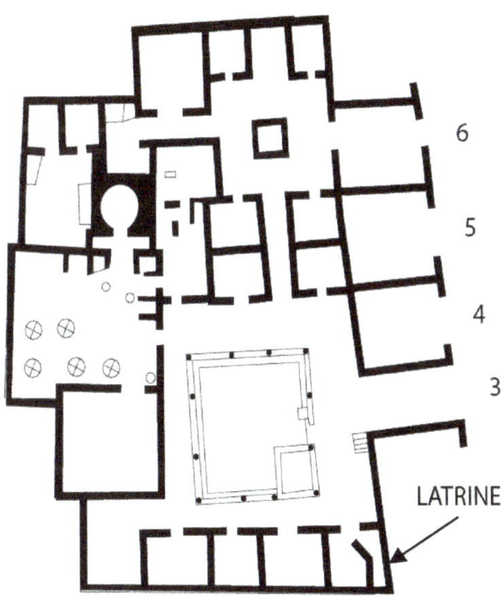

VII. 2. 3

VII. 2. 9

VII. 2. 12 Down pipe in staircase

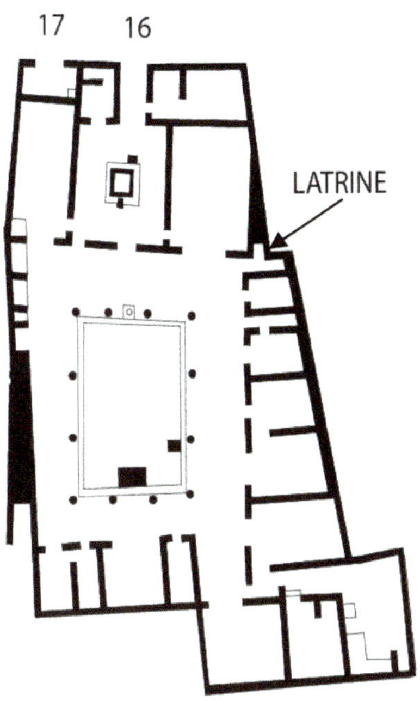

VII. 2. 16 Latrine

VII. 2. 18

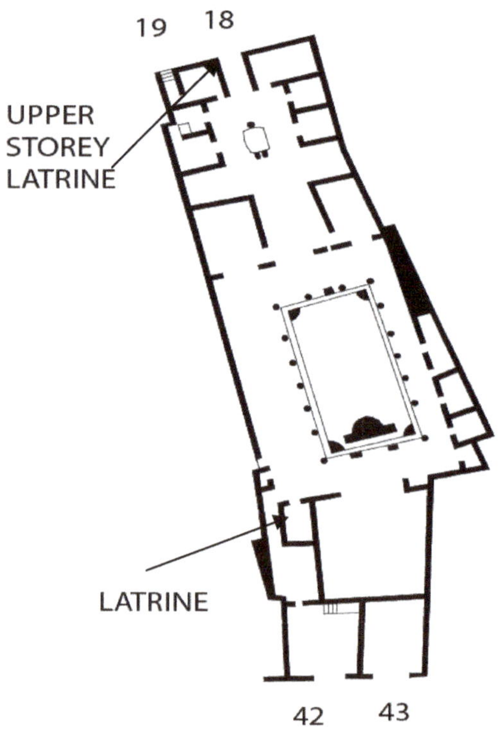

VII. 2. 18

VII. 2. 20

VII. 2. 20 Upper storey latrine and down pipe

VII. 2. 23 Down pipe: South wall room g

VII.2.24 Downpipe

VII. 2. 27, 29-31, 35 Casa di Mercurio

VII.2.34 Down pipe

VII.2.36 Down pipe

VII. 2. 37-39

VII. 2. 37, 38 Down pipe east wall
room b

VII. 2. 40

VII. 2. 42 Downpipe

VII. 2. 45 Latrine

VII. 2. 46 Downpipe

VII. 2. 46

VII. 2.51/52: Downpipe in outer wall
between #51&#52

VII.2.51 Latrine and down pipe scar: room s

REGIO VII.3

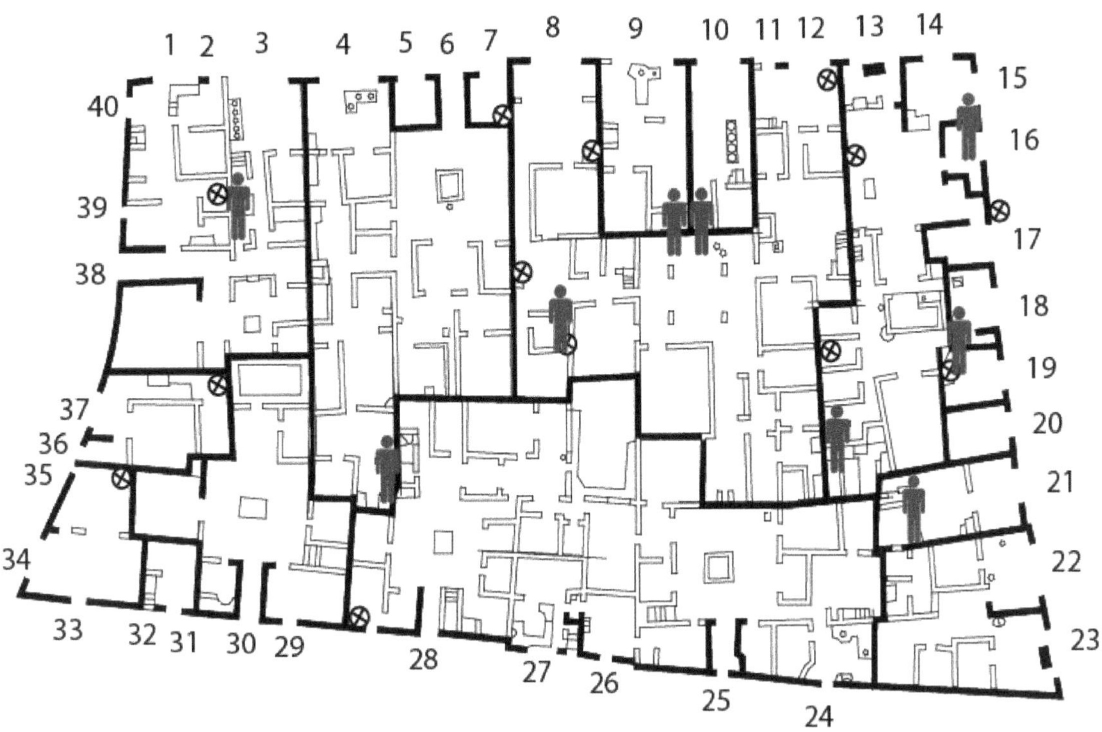

Seven small shops, five medium and five large houses can be seen in this insula. There is a sufficiency of latrines and some down pipes indicating upper storey latrines.

VII. 3. 1-3, 38-40 *Workshop of C. Memmius*

LATRINE room d

VII. 3. 4-7 *Workshop with shop* etc.

Latrine in 4 (Eschebach p265:Fiorelli p85)

UPPER STORE LATRINE

DOWN PIPE (not shown on plan)

VII. 3. 8 *Workshop and shop*

TWO DOWN PIPES

Latrine (Fiorelli p86)

VII. 3. 9 *'Popina'*

LATRINE (Eschebach p266:Fiorelli p86)

VII. 3. 10 *Shop*

LATRINE (Eschebach p267)

VII. 3. 11-15, 17 *Large workshop with shop*

VII. 3. 13-15, 17 *Casa del doppio Larario*

LATRINE (Eschebach p267:Fiorelli p86)

VII. 3. 16 *Shop*

LATRINE with DOWN PIPE in floor

VII 3. 18 *Shop*

LATRINE (Fiorelli p86)

VII. 3. 19 *Shop of Secundus*

DOWN PIPE commented upon by Fiorelli, p86

VII. 3. 20 *Shop of Rubellius*

Nothing

VII. 3. 21 *Shop*

LATRINE (Eschebach p269:Fiorelli p87)

VII. 3. 22, 23 *Workshop with shop*

Latrine in 23 (Eschebach p 269)

Nothing

VII. 3. 24, 25 *Workshop*

Latrine in 24 (Eschebach p 269)

NO ACCESS (BUILDERS)

VII. 3. 26, 28 *Café bar and brothel*

DOWN PIPE

VII. 3. 29 *House of M Spurius Mensor*

Two DOWN PIPES

Latrine (Eschebach p 270)

NO ACCESS

VII. 3. 30 *House of anonymous magistrate*

Latrine (Eschebach p 270)

NO ACCESS (BUILDERS)

VII. 3. 31, 32 *Shop*

Latrine ((Eschebach p 271)

Nothing

VII. 3. 33-35 *Small workshop*

DOWN PIPE

VII. 3. 36, 37 *Workshop*

Nothing

VII.3. 38-40

DOWN PIPE

VII.3.1-3,38-40 Latrine

VII. 3. 4 Upper storey latrine room p

VII. 3. 3

VII.3.5,6 Down pipe

VII 3. 6

VII. 3. 8

VII. 3. 8 Down pipe in masonry

VII. 3. 10

VII. 3. 12 room e

VII. 3. 13 room a

VII. 3. 16

VII. 3. 15, 17 Latrine with down pipe in floor

VII. 3. 17

VII. 3. 18

VII. 3. 19

VII. 3. 21

VII. 3.26, 28

VII. 3. 29 Outer wall: Two down pipes

VII. 3. 39

VII. 3. 35 Down pipe

REGIO VII.4

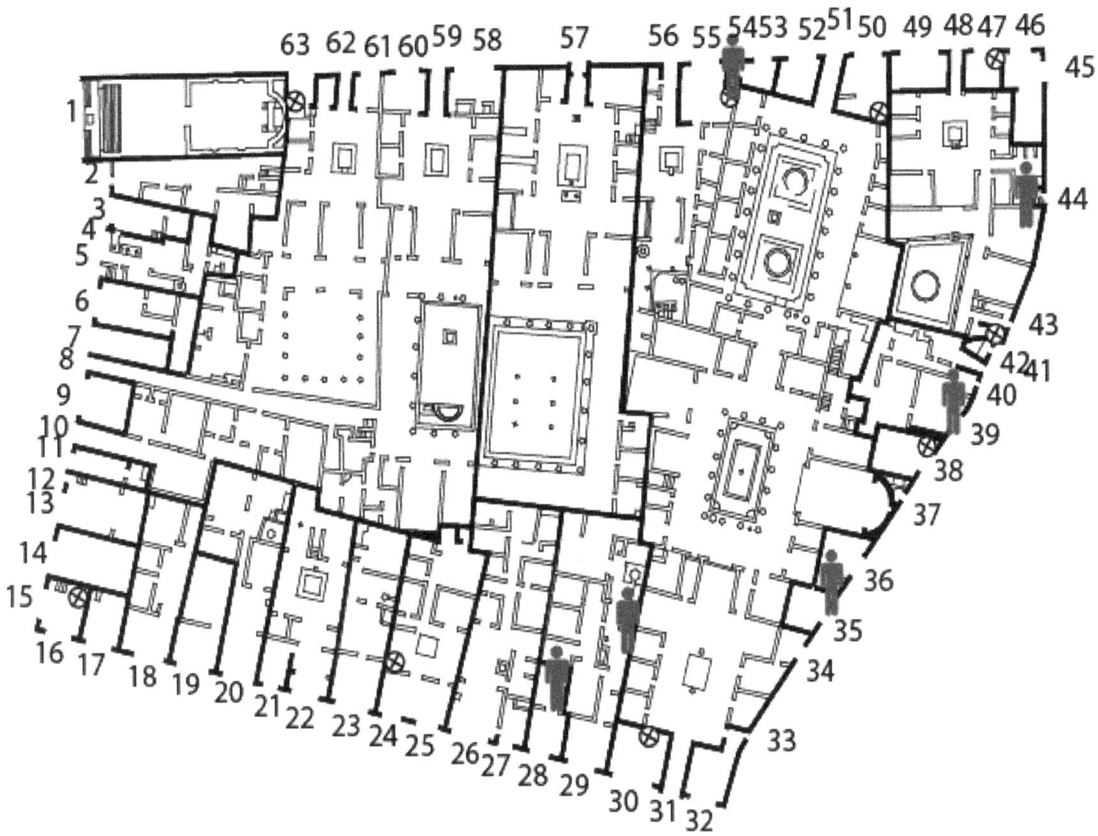

A large insula containing thirty one properties but dominated by three large houses, and a temple at the northwest corner. Eschebach lists a number of latrines, the existence of some of which has not been verified in this survey.

VII. 4.1, 2 *Temple of Fortuna Augusta*

Latrine in 2 (Eschebach p272: Fiorelli p89)

NO ACCESS

VII. 4. 3 *Shop*

Nothing

VII. 4. 4 *Café bar*

Latrine (Eschebach p273 Fiorelli p89)

Nothing

VII. 4. 6, 7 *Large shop*

Nothing

VII. 4. 8, 10, 59, 62 *House of Bacco*

Modern offices

(Eschebach p273, 274)

VII. 4. 9 *Shop*

Nothing

VII. 4. 11 *Shop*

Nothing

VII. 4. 12, 13 *Shop*

Nothing

VII. 4. 14 *Shop of Aurelius*

Nothing

VII. 4. 15, 16 *Wine bar of Samellius*

DOWN PIPE in Northeast corner

VII. 4. 17 *Shop*

Nothing

VII. 4. 18 *Small house*

Latrine (Eschebach p275: Fiorelli p90)

NO ACCESS

VII. 4. 19 *Shop*

Nothing

VII. 4. 20 *House of a wine merchant*

Latrine (Eschebach p276: Fiorelli p90)

Nothing

VII. 4. 21, 22 *Workshop of Trebbius*

Nothing

VII. 4. 23

Latrine (Eschebach p277)

Nothing

VII. 4. 24, 25, 59

Nothing

VII. 4. 26, 27 *Workshop*

Large latrine in 26 (Eschebach p278 Fiorelli p90)

Nothing

VII. 4. 28 *Shop*

Latrine (Eschebach p278 Fiorelli p90)

Possible in room at rear.

VII. 4. 29, 57 *Shop to rear of house*

See 57

VII. 4. 30 *Shop*

DOWNPIPE

VII. 4. 31-33, 50, 51 *House of the Coloured Capitals*

31 LATRINE (Fiorelli p91)

DOWN PIPE room 10

VII. 4. 34 *Shop*

Latrine (Eschebach p280)

VII. 4. 35 *Shop*

VII. 4. 36 *Shop*

LATRINE under stairs

VII. 4. 37 *Shop ?cella meretricia*

VII. 4. 38 *Very large shop*

DOWN PIPE (Fiorelli, p92)

VII. 4. 39-41 ***Dyehouse and shop***

Nothing

41 LATRINE (Eschebach p281 Fiorelli p92)

VII. 4. 42 Cella meretricia

Latrine (Eschebach p281: Fiorelli p92, public latrine)

VII. 4. 43, 48 ***Casa dell Caccia antica***

48 LATRINE room 21. One large pedestal and one narrow pedestal. There is an upper floor down pipe into the latrine which is flushed with overflow water from the peristyle (Allison & Sear 2002,51) (Eschebach p281: Fiorelli p92)

VII. 4. 44 Latrine from street

(Eschebach p281)

No evidence

VII. 4. 45-47 ***Workshop***

DOWN PIPE in corner

VII. 4. 49 ***Shop***

VII.4.50 ***Shop***

DOWN P IPE

VII. 4. 52 ***Shop***

Latrine (Eschebach p282)

Nothing

VII. 4. 53, 54 ***Shop***

53 LATRINE (Fiorelli p92)

VII. 4. 55 ***Large shop***

Nothing

VII. 4. 56 ***Casa del Granduca di Toscana (Casa della Fontana)***

LATRINE, room 18 (Gierow 1994, 38) (Eschebach p283: Fiorelli p93)

VII. 4. 57, 29 ***House of the figured Capitals***

Eschebach latrine in 29

No latrine shown in Gierow (ibid)

VII. 4. 58-63, 8 Casa della Parete Nera Casa delle Forme di Creta

59 LATRINE room u (Gierow 2000,54)

Latrines in 59 & 62 (Eschebach p284) (Fiorelli p 94)

VII.4.15,16 Downpipe in NE corner

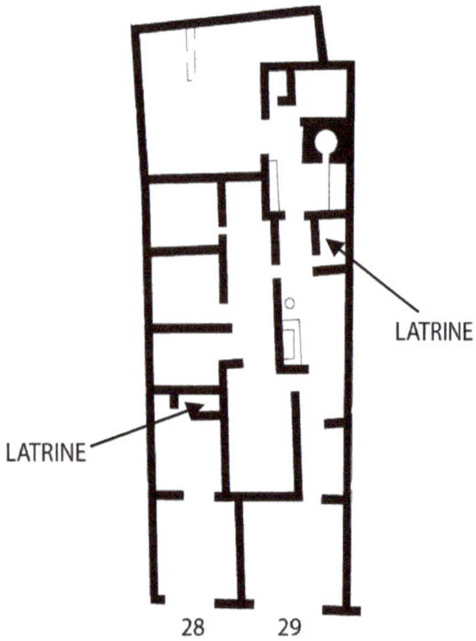

VII.4.28

VII.4.29

VII.4.31,51 Down pipe

VII.4.30 Downpipe excavation

VII.4.36

VII.4.38

VII.4.43,48

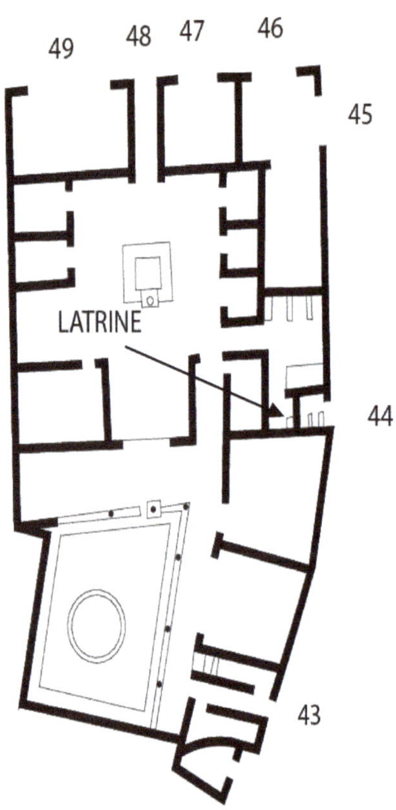

VII.4.48

VII.4.47 Down pipe

VII.4.50 Double Down pipe

VII.4.51 (2005)

The room of the latrine in the House of the Coloured Capitals photographed in 2005 and then in 2007. Looking at the earlier picture there is little to show that it was the site of the latrine

VII.4.51 (2007)

VII.4.54

VII.4.56

VII.4.59

VII.4.60/61 Down pipe interior of wall between #60 and #61

VII.4.63 Down pipe

REGIO VII 6

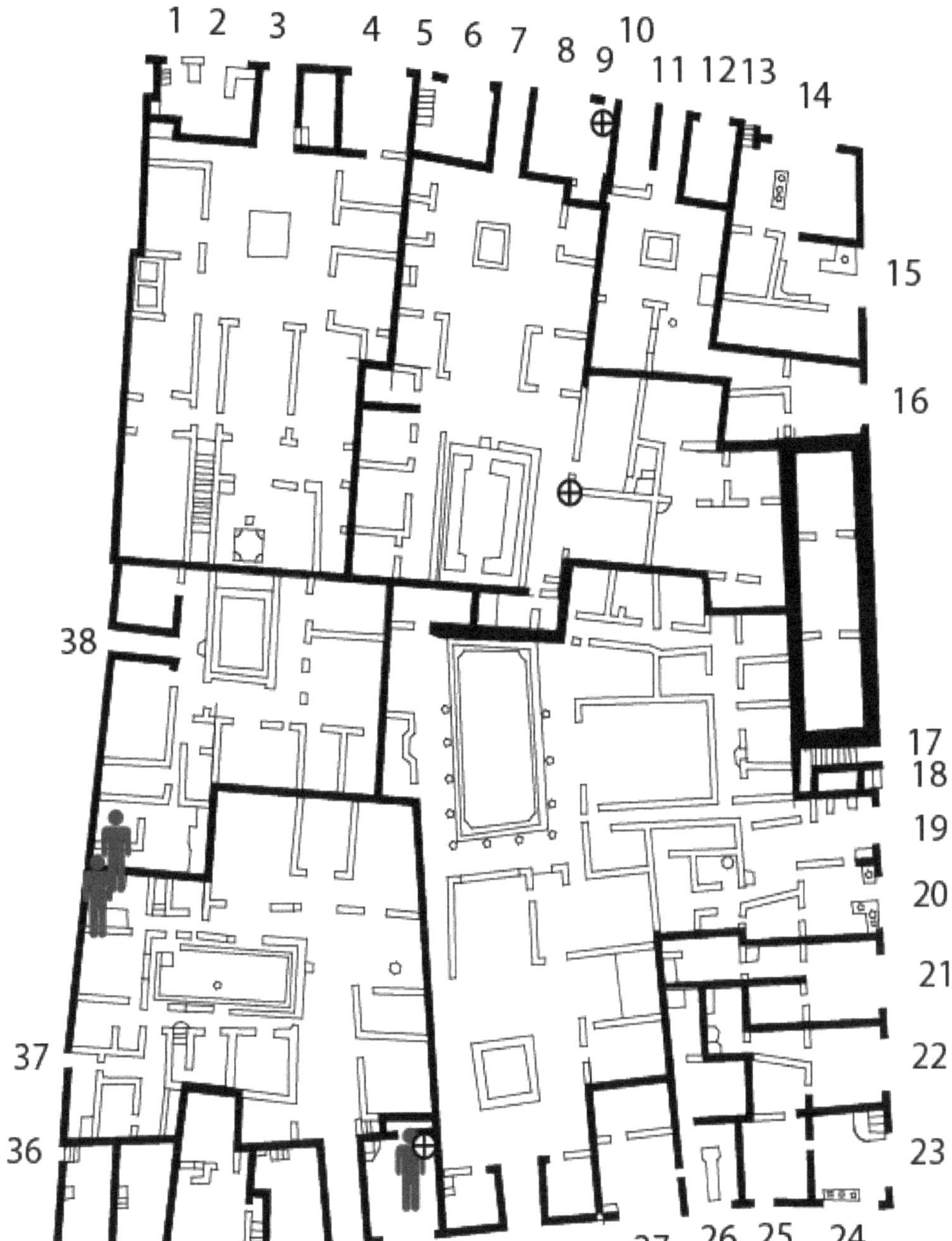

A very badly damaged area of the city. Hardly any evidence of latrines or downpipes

VII. 6. 1, 2 ***Large shop or bar***

Nothing

VII. 6. 3, 4 ***House of M. Spurius Saturninus***

Nothing

VII. 6. 5, 6 ***Shop***

Nothing

VII. 6. 7 ***Private house***

DOWN PIPE

VII. 6. 8, 9 ***Shop***

DOWN PIPE

VII. 6. 10, 11, 16

Nothing

VII. 6. 12 ***Small shop***

Nothing

VII. 6. 13-15 ***Wine shop of Epidius Sabinus***

Nothing

VII. 6. 17, 18 Water reservoir

Nothing

VII. 6. 19, 20, 28 ***Dyehouse***

Nothing

VII. 6. 21 ***House***

Latrine (Eschebach p295)

Nothing

VII. 6. 22-25 ***Workshop with bar***

Nothing found

VII. 6. 26, 27 ***Bar of C. Marcellus***

Latrine in 26 (Eschebach p296)

Nothing

VII. 6. 28, 19, 20 ***House of Secundus Tyrannus Fortunatus***

Latrine in 28 (Eschebach p296)

Nothing

VII. 6. 29 ***Workplace and shop***

DOWN PIPE

POSSIBLE LATRINE under stairs

VII. 6. 30, 37. ***House of Petutius Quintio***

LATRINE in north western corner of the property (room 47) (Eschebach p297)

With accompanying DOWN PIPE from upper storey

VII. 6. 31 ***Shop***

Nothing found

VII. 6. 32, 33 ***Shop***

Nothing found

VII. 6. 34-36 ***Brothel of Venus***

Nothing found

VII. 6. 38 ***House of Cipius Pamphilus Felix***

Latrine (Eschebach p298)

POSSIBLE LATRINE in room 35

VII. 6. 9 Down pipe

VII. 6. 29 Down pipe

VII. 6. 7 Down pipe

VII. 6. 29

VII. 6. 30, 37

VII. 6. 30, 37 After cleaning: 2007

VII.6.38
Possible latrine in room 35

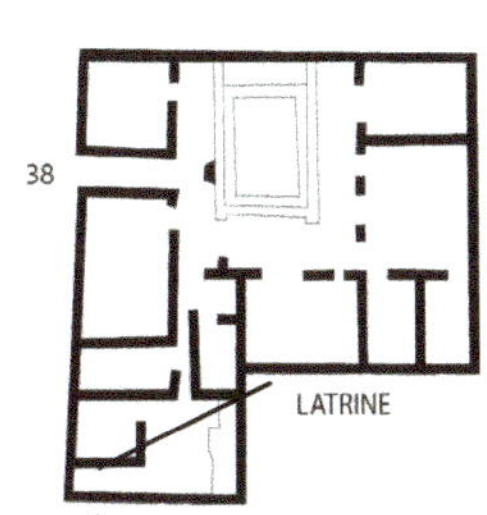

REGIO VII. 7

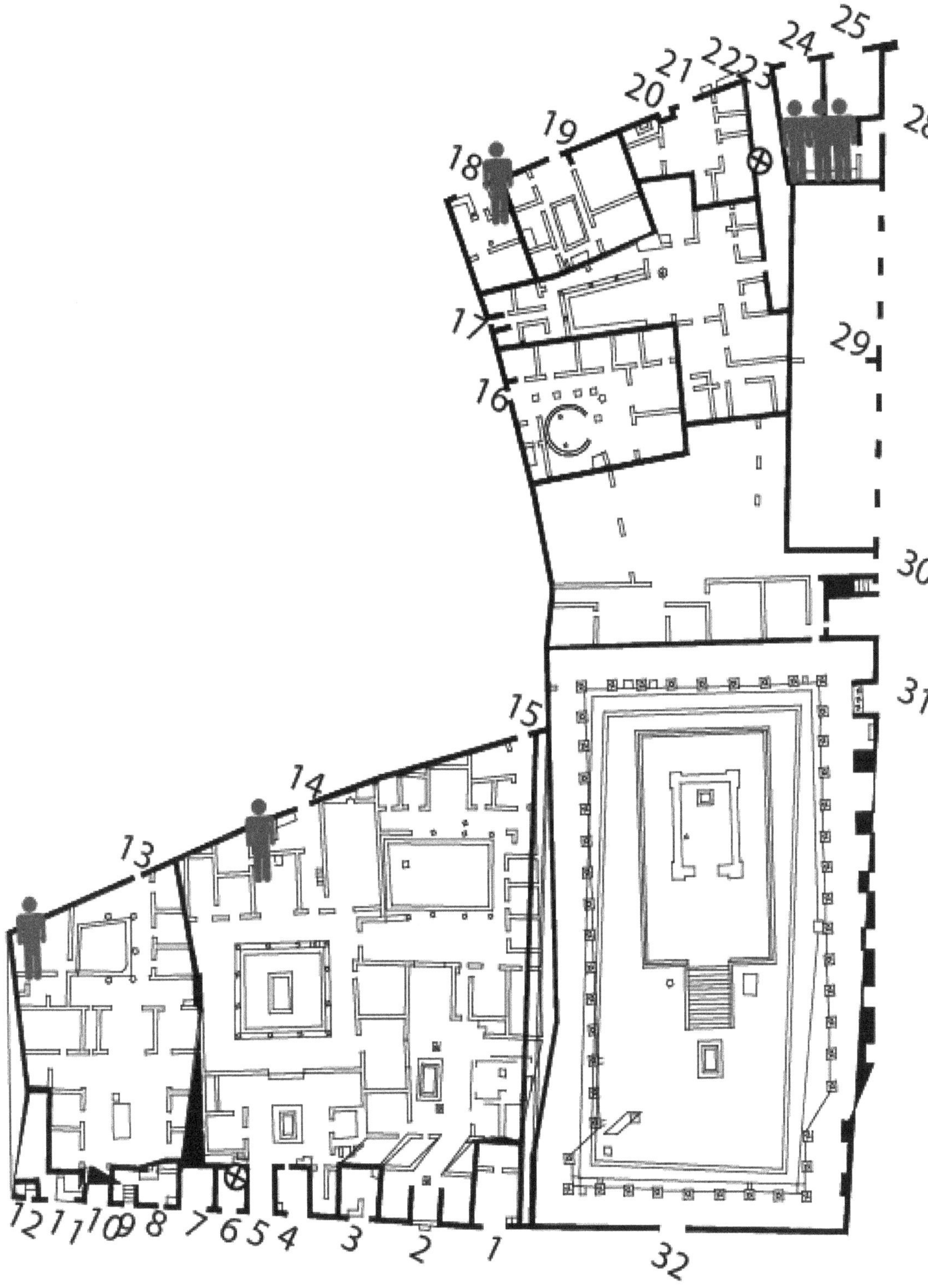

An interesting insula to the south west of the Temple of Apollo. Two large houses and five shops. Some evidence of upper storey latrines.

VII. 7. 1 ***Shop***

VII. 7. 2, 5, 14, 15 *House of Cissonius* ***Secundus***

Latrine in #5 (Eschebach p299: Fiorelli p98)

5 POSSIBLE LATRINE no photograph

VII. 7. 3 ***Shop***

Nothing

VII. 7. 6 ***Shop***

Down pipe scar in wall

VII. 7. 7 ***Shop***

Nothing

VII. 7. 8, 9 ***Café/bar***

PROBABLE LATRINE

VII. 7. 10, 13 ***House of Fabius H.***

13 PROBABLE LATRINE

VII. 7. 11, 12 ***Workshop and shop of Furius***

Latrine in 11 not confirmed (Eschebach p301: Fiorelli p99)

VII. 7. 16 ***House of veteran Julianus***

Latrine in 16 not confirmed (Eschebach p302: Fiorelli p100)

VII. 7. 18 ***Bar of L Numinius with brothel***

LATRINE in north east corner room 5

VII. 7. 19 ***House of Jeweller***

Nothing

VII. 7. 20, 1 ***House of the altar of Jupiter***

Nothing

VII. 7. 23, 17 ***House of Julius Primigenius***

Latrine in 17 not confirmed (Eschebach p 303)

DOWN PIPE in outer wall between #22 and #23

VII. 7. 24 ***Shop***

Nothing

VII. 7. 28 ***Public Latrine*** (Eschebach p304: Fiorelli p101)

VII. 7. 29 ***Bargain shop***

Nothing

VII. 7. 31, 32, 34, 35 ***Temple of Apollo***

Nothing

VII. 7. 6 Down pipe scar

VII. 7. 8, 9 Probable latrine

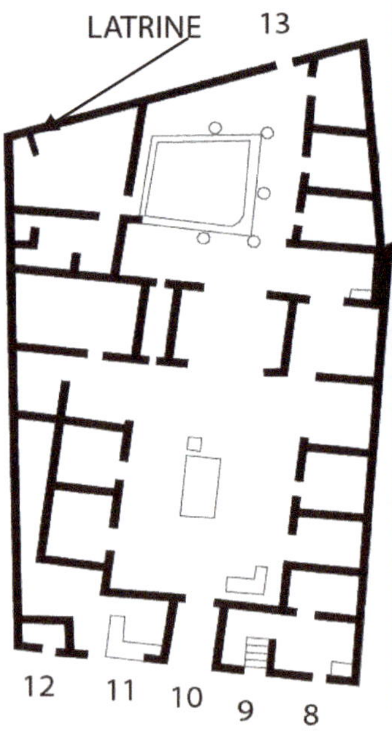

VII. 7. 18 Latrine

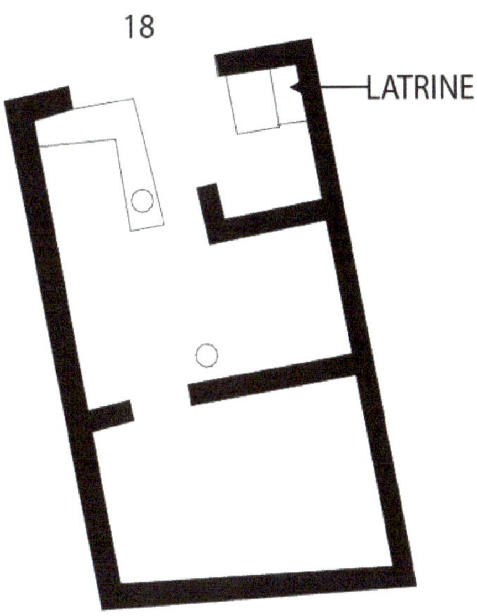

VII. 7. 22, 23 Down pipe in outer wall

REGIO VII 9

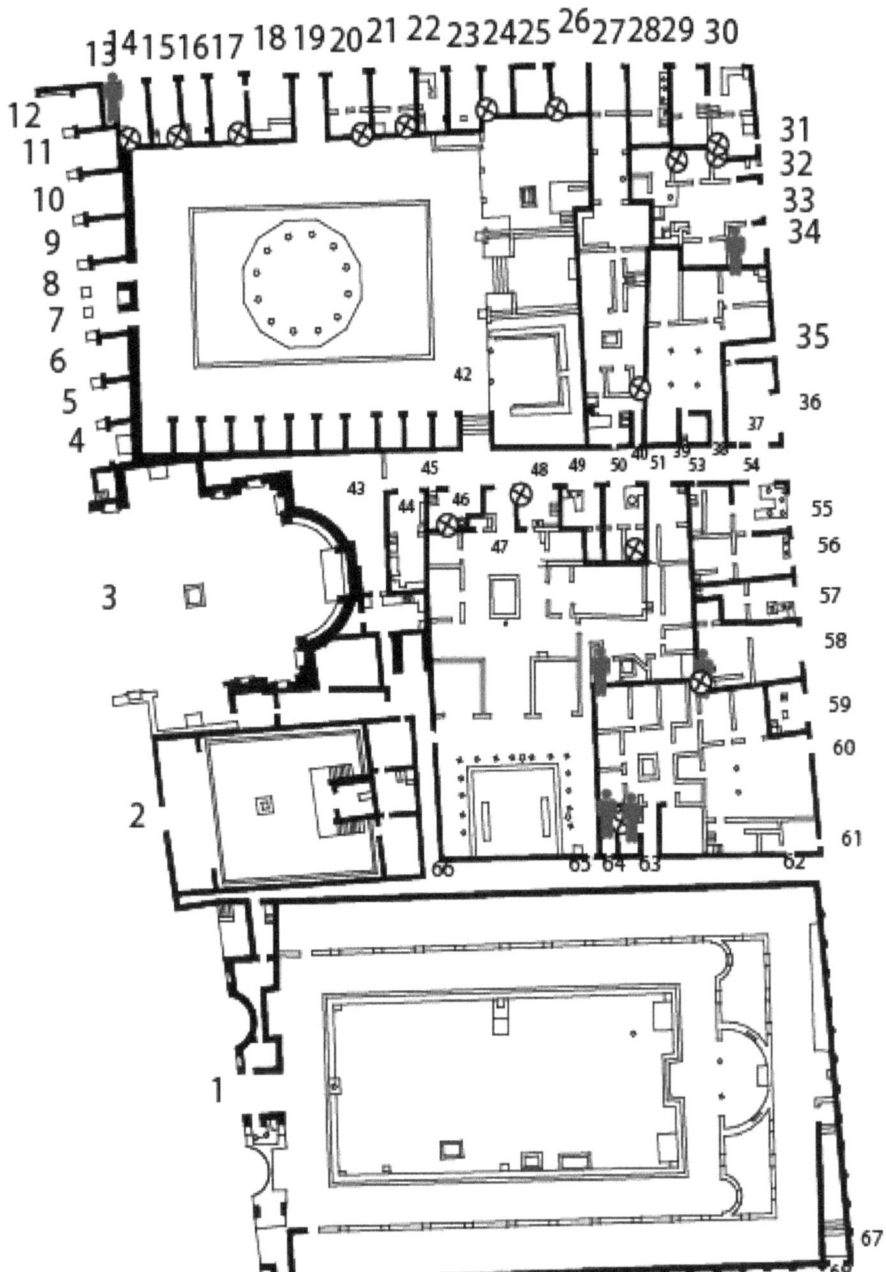

A very absorbing area with properties hidden behind the buildings to the east of the forum. The properties lining the Via degli Augustali appear to have down pipes on a line. Many of Escherbach's latrines were not found.

The chronological development of the latrines in the House of the Fisherman is particularly interesting.

VII. 9. 1, 43, 66-68 *Eumachia Building*

1 Gastrum urinarium

VII. 9. 2, 43, 66 *Temple of Vespasian*

Latrine in 66 (Eschebach p308)

Nothing

VII. 9. 5 *Tavern*

Nothing

VII. 9. 6 *Tavern*

Nothing

VII. 9. 7, 8, 19, 42 *Macellum*

Nothing

VII. 9. 10 *Shop*

Nothing

VII. 9. 11 *Tavern of Cornelius*

Nothing

VII. 9. 12 *Shop of Quintus*

Nothing

VII. 9. 13 *Sodalicium*

LATRINE

VII. 9. 14 *Shop*

DOWN PIPE

VII. 9. 15 *Shop of Similis*

Latrine (Eschebach p312)

DOWN PIPE

VII. 9. 16 *Shop of Piaulis*

Nothing

VII. 9. 17, 18 *Shops*

DOWNPIPE in 17

Latrine in 18 (Eschebach p312)

VII. 9. 20 *Shop*

Latrine (Eschebach p312)

DOWN PIPE

VII. 9. 21 *Shop*

DOWN PIPE

VII. 9. 22 *Café*

Niche for LATRINE not evident

VII. 9. 23 *Shop*

Nothing

VII. 9. 24 *Small shop*

DOWN PIPE

VII. 9. 25 *Shop*

Nothing

VII. 9. 26 *Shop*

DOWN PIPE

VII. 9. 27, 40, 41 Dyehouse

Latrine (Eschebach p315:Fiorelli p106)

Nothing

VII. 9. 28 *Shop*

Latrine (Eschebach p314: Fiorelli p106

Nothing

VII. 9. 29-34 *Café and bar of Donatus and Verpus*

DOWNPIPE in 29 room 2

DOWN PIPE in 31 room 3

DOWN PIPE in 33 room 4

Latrine in 33 (Eschebach p314: Fiorelli p106)

LATRINE in 34

VII. 9. 35-37 *Workhouse*

Latrine (Eschebach p315:Fiorelli p106)

Nothing

VII. 9. 38.39 *Small house*

Latrine in 38 (Eschebach p315)

Nothing LATRINE

VII. 9. 40, 41, 27 ***Dyehouse***

Nothing

VII. 9. 44 ***Dyehouse***

Nothing

VII. 9. 45, 46 ***Shop***

Nothing

VII. 9. 47, 48, 51, 65 ***House of a Flamen***

latrine in 47 (Eschebach p316: Fiorelli p107)

Not evident

LATRINE in 51 with DOWN PIPE (Eschebach p316: Fiorelli p107)

VII. 9. 49 ***Shop***

Nothing found

VII. 9. 50 ***Larger shop***

DOWNPIPE

Eschebach ?latrine not evident

VII. 9. 53-56 ***House, bar and shop***

Latrine in 53((Eschebach p317: Fiorelli p107)

Nothing

VII. 9. 57 Café

Latrine (Eschebach p317)

Nothing

VII. 9. 58 ***Large shop***

LATRINE AND DOWNPIPE (Eschebach p318: Fiorelli p107)

VII. 9. 59 Shop.

Nothing

VII. 9. 60, 63 ***House of the fisherman***

LATRINE and DOWNPIPE in 63

VII. 9. 64 ***Stairs to upper storey***

VII. 9. 13

VII.9.14

VII.9 15 Down pipe

VII. 9. 17 Downpipe

VII.9.20 Down pipe

VII.9.21

VII.9.24 Down pipe

VII.9.25 Down pipe

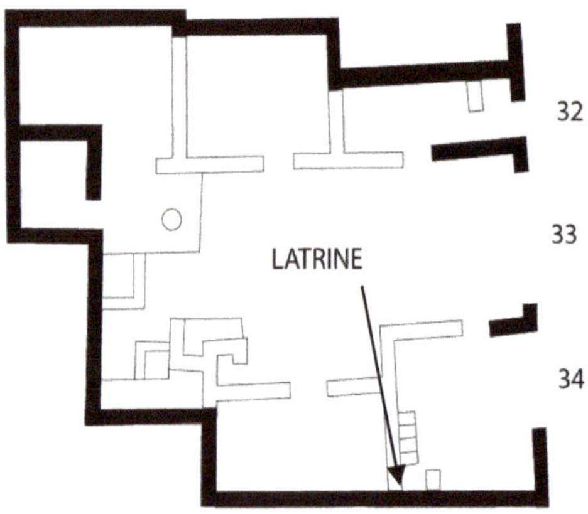

VII. 9. 33 Down pipe NW corner
room 4

VII. 9. 33 Two down pipes: room 4

VII. 9. 34

VII.9.40 Down pipe

VII. 9. 45 Down pipe

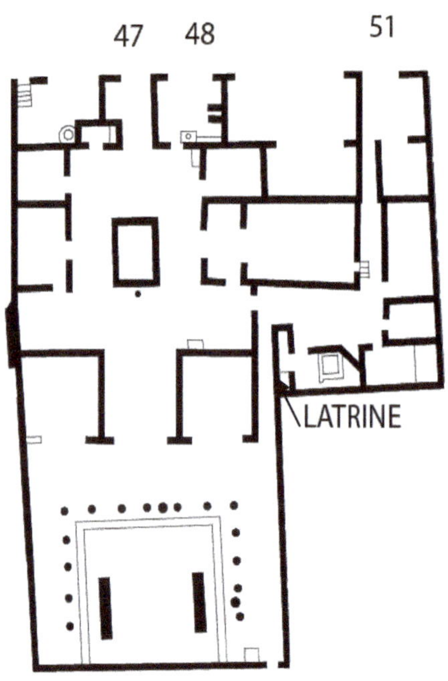

VII. 9. 47, 65

VII. 9. 48 Down pipe

VII. 9. 50 Downpipe

VII. 9. 58 Latrine and down pipe

VII.9.63 Down pipe

VII. 9. 63 Latrine with down pipe in wall

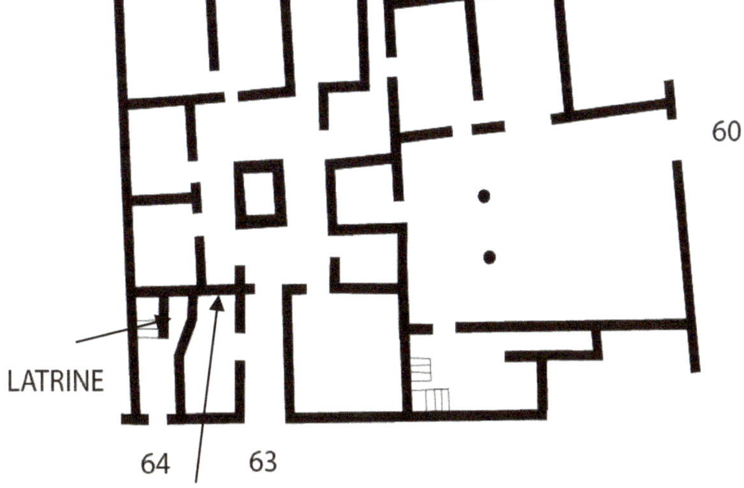

VII. 9. 64 Latrine by staircase

REGIO VII.10

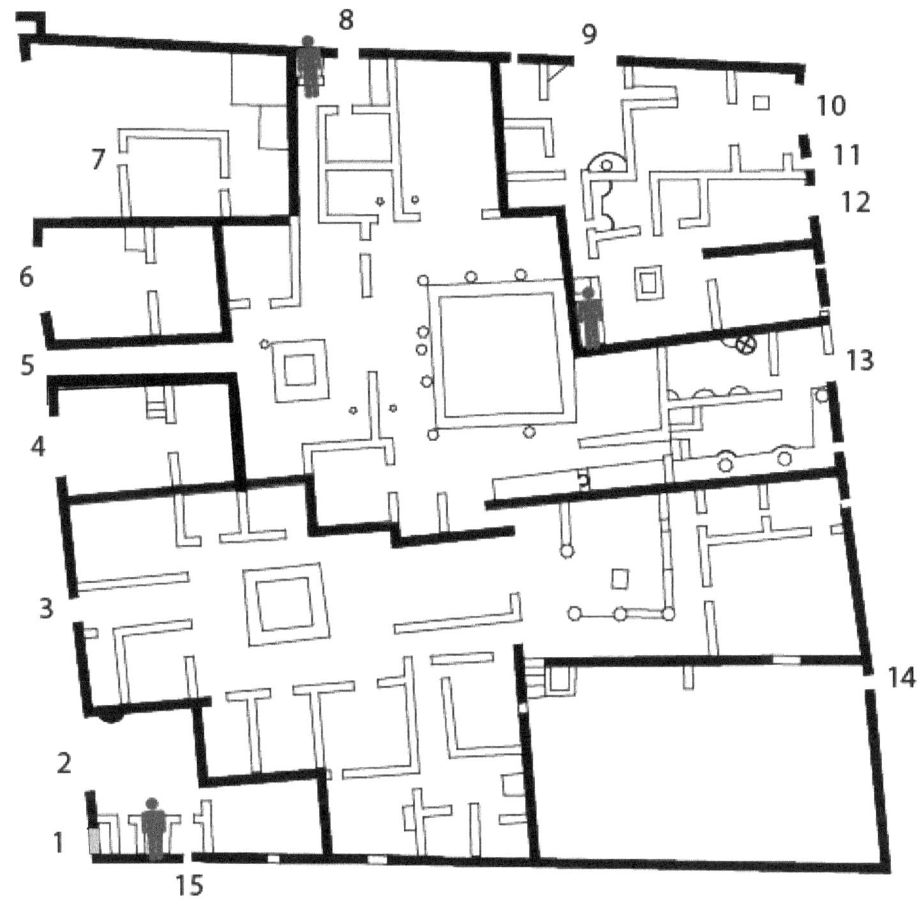

Six properties and a garden, three latrines, one down pipe.

VII. 10. 1, 2, 15 *Workshop*

On the south side of the property is a small area filled with rubble which may be the location of the latrine

Latrine in 15 (Eschebach p319: Fiorelli p108)

VII. 10. 3, 14 *Casa della Caccia nuova*

(Haüser. Sear and Allison) (Eschebach p320: Fiorelli p108)

Latrine not evident

VII.10.4 Shop

Nothing found

VII. 10. 5, 8, 13 *Workshop*

LATRINE Probable by doorway #8

DOWN PIPE in #13

VII. 10. 6 *Shop*

Nothing found

VII. 10. 7 *Shop*

Nothing found

VII. 10. 9, 12 *Small workshop*

LATRINE

in southwest corner of property, off room g

Allegedly there is a second latrine in this property (Escherbach p321)

This is seen in the illustration but not confirmed

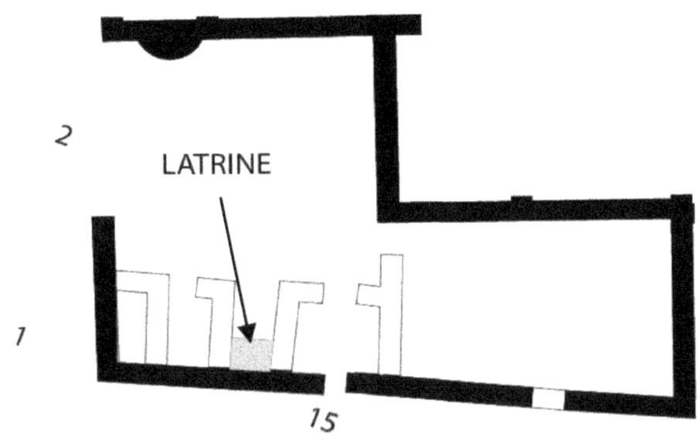

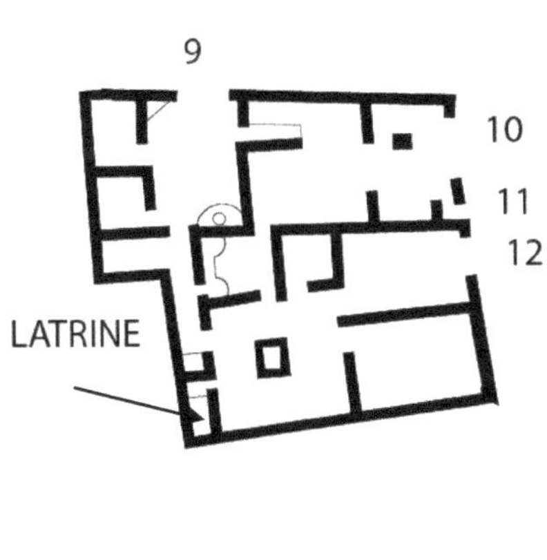

VII. 10. 9, 12

?VII. 10. 12 (note wrongly labeled board). Rubble filled area for possible second latrine in property

VII. 10. 13 Down pipe: North wall

REGIO VII.11

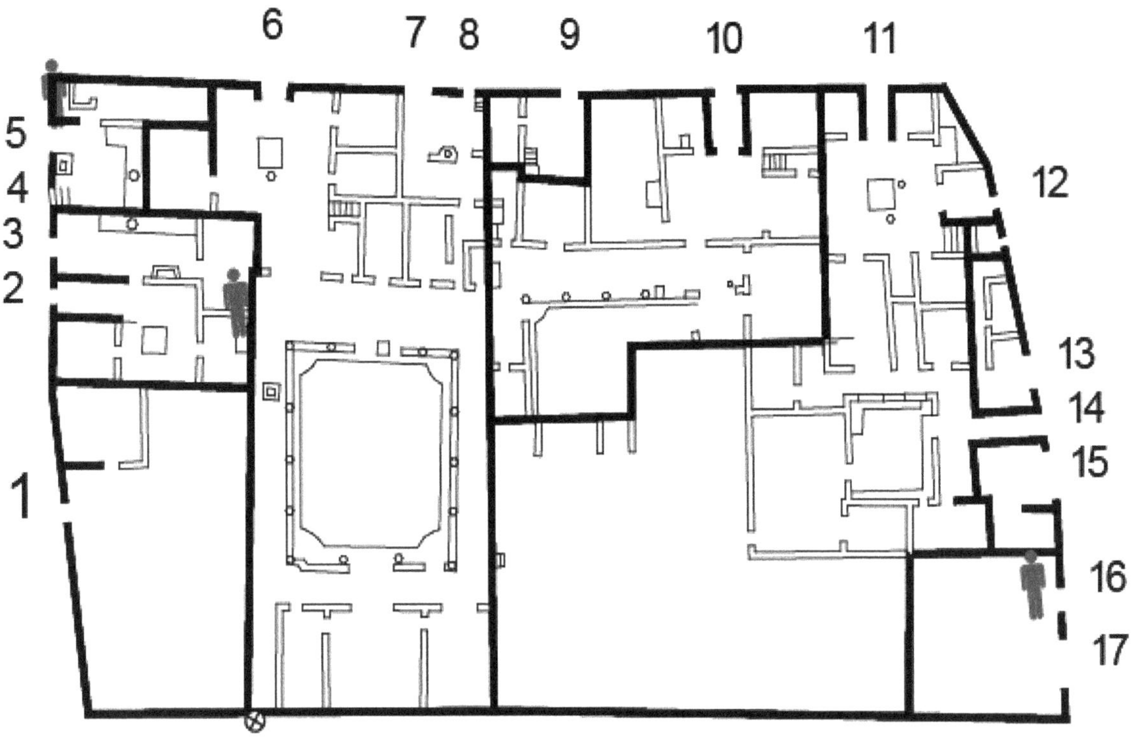

Three large houses two smaller ones and two open spaces.

VII 11, 1 *Garden*

Nothing found

VII 11, 2.3. *Laundry*

LATRINE in 2

VII 11, 4.5 Laundry

LATRINE in 5 (Fiorelli p109)

VII 11, 6-*8 Inn*

6 Eschebach ?Latrine not confirmed

VII 11, 9.10 *Workshop*

NO ACCESSS

Latrine (Fiorelli p109)

VII 11, 11.12.14 *Inn of the Christians*

Latrine (Eschebach p324: Fiorelli p110)

Nothing

VII 11,13 *Bar of Felix*

Latrine (Eschebach p324: Fiorelli p110)

not confirmed

VII 11, 15 *Shop*

Eschebach Latrine

Nothing

VII 11, 16.17 *Workshop*

LATRINE

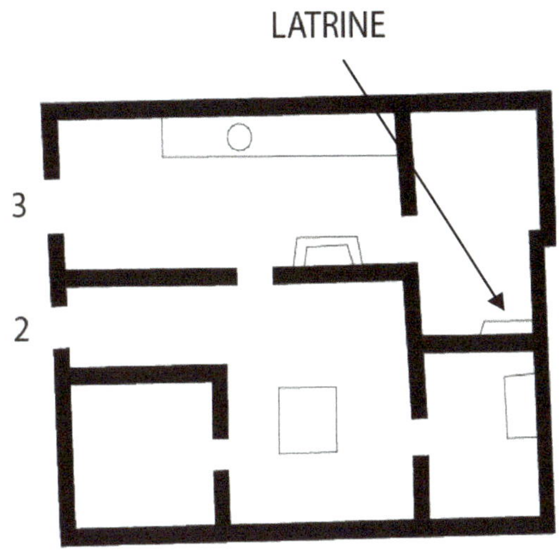

VII.11.2,3

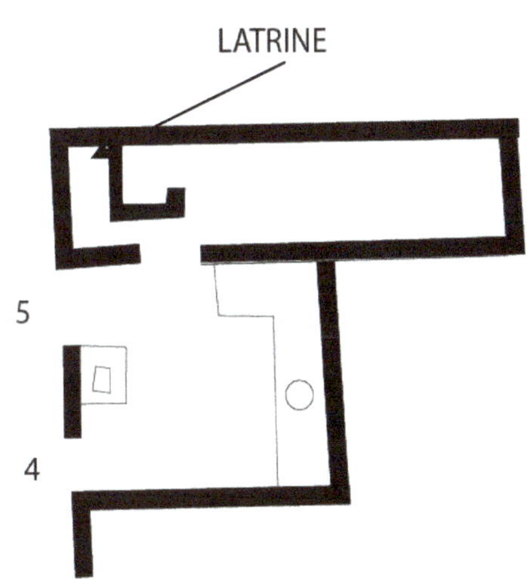

VII.11.4,5

VII.11.6 Downpipe in external wall of room k

VII.11.16,17

REGIO VII.12

An *insula* with fifteen properties, small houses and shops. Two upper storey latrines.

VII. 12. 1, 2, 36 *Pistrinum of Donatus*

Latrine (Eschebach p325)

Nothing

VII. 12. 3, 4 *House of L. Caecilius*

Capella

Latrine in 4 (Eschebach p326)

DOWN PIPE in room a

LATRINE in room n

VII. 12. 5 *Shop of Rutellius*

LATRINE in room b (Eschebach p326: Fiorelli p110)

VII. 12. 6 Shop

Nothing

VII. 12. 7 Pistrinum

Nothing

VII. 12. 10 *Shop of Festus*

Latrine (Eschebach p327: Fiorelli p111)

Possible LATRINE

VII. 12. 11 *Casa del Forno*

LATRINE room d (Eschebach p327: Fiorelli p111)

VII. 12. 12 Workshop

Latrine (Eschebach p3286: Fiorelli p111)

Nothing

VII. 12. 13 *Bakery and Bar of Sabinus*

LATRINE and separate DOWN PIPE (Eschebach p328)

VII.12.14 *Schoolhouse?*

Latrine (Eschebach p328: Fiorelli p111)

UPPER STOREY LATRINE room f with TWO DOWN PIPES

VII. 12. 15, 16 *Bar Of Paris*

Latrine (Eschebach p329: Fiorelli p111)

Possible

VII. 12. 17, 21 *Casa Di Narciso*

Latrine in 17((Eschebach p329: Fiorelli p112)

UPPER STOREY LATRINE in 21

VII. 12. 18, 20 *Brothel of Africanus*

LATRINE

Latrine by entrance to 19 (Eschebach p330: Fiorelli p112)

LATRINE inside entrance to 20 which is staircase to upper storey of brothel

VII. 12. 22, 23 *Casa del Camillo*

LATRINE in 23 room O (Eschebach p330: Fiorelli p112)

DOWN PIPE in room b

VII. 12. 25-27 *House of Cornelius Diadumenus*

DOWN PIPE

Curtis quotes 'a latrine and a kitchen for workmen'(Curtis p19)

VII. 12. 28 *Casa del Balcone pensile*

Latrine (Eschebach p331: Fiorelli p114)

NO ACCESS: Building work

VII. 12. 29 Public Latrine

Latrine not evident (Eschebach p331)

VII. 12. 30-32 Small workshop

Latrine in 31 (Eschebach p332)

LATRINE

VII. 12. 33 Cella meretricia

Nothing

VII. 12. 34,35 Inn

Latrines in 34 & 35 (Eschebach p332: Fiorelli p114: Packer shows latrines in 34 room 15 and 35 room 9)

DOWN PIPE

VII.12. 4 Room a down pipe

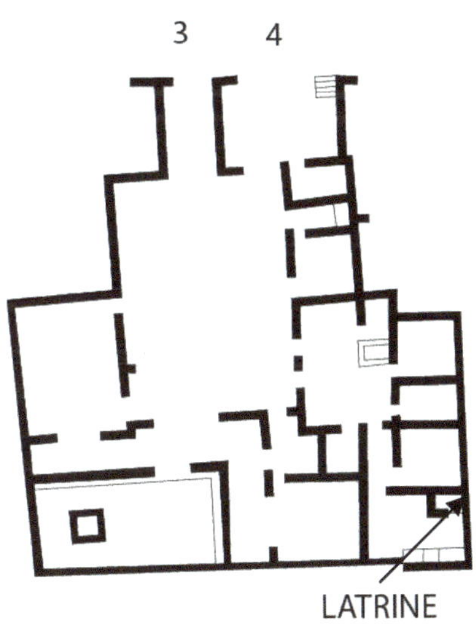

VII. 12. 3, 4 Latrine

VII. 12. 10 Possible latrine

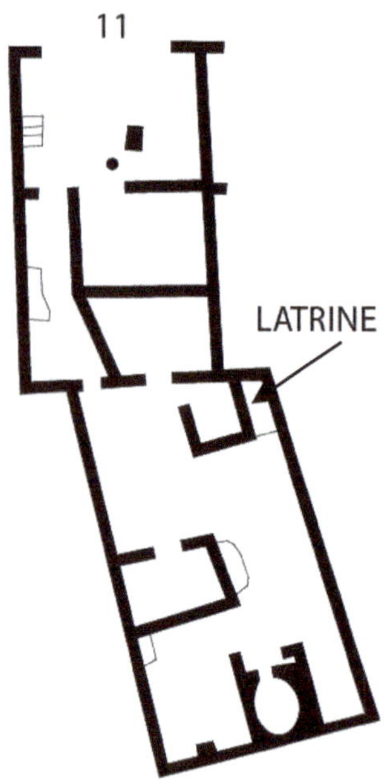

VII. 12. 11 Latrine in room d

VII. 12. 13 Down pipe

VII. 12. 13 Latrine

VII. 12. 14 Down Pipe

VII. 12. 15, 16

VII.12.19 Latrine in brothel of Africanus

VII.12.20 Latrine in the rear staircase of the brothel of Africanus

VII. 12. 21 Room h western wall

VII. **12. 22 room O**

VII. 12. 25-27 Down pipe

VII. 12. 22 room b

VII. 12. 22 room O

VII. 12. 30-32

VII. 12. 35 Down pipe

REGIO VII.13

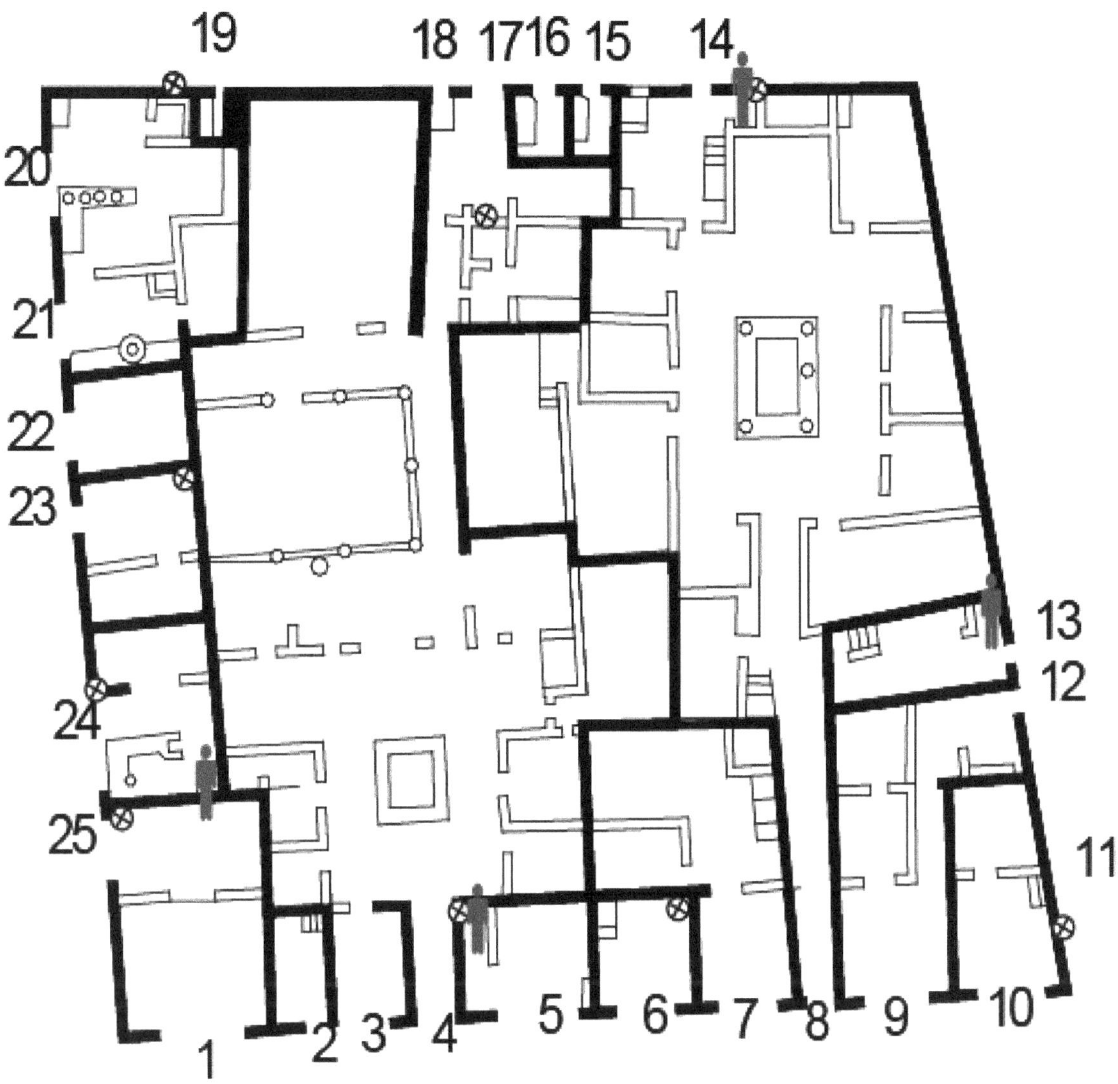

One of the most interesting latrines in Pompeii (VII.13.14) is to be found in this insula.

VII. 13. 1, 25 ***Shop*** **[See VII.13.25]**

VII. 13. 2 ***Small shop***

VII. 13. 3, 4, 16-18 ***Casa di Ganymede***

LATRINE in 4

(Eschebach p333: Fiorelli 115)

VII. 13. 5 ***Shop***

Latrine (Eschebach p333)

DOWN PIPE & LATRINE

VII. 13. 6 ***Shop***

Latrine (Eschebach p334: Fiorelli 115)

DOWN PIPE

VII. 13. 7 ***Small Workshop***

Latrine (Fiorelli p115)

VII. 13. 9, 12 Small workplace\ shop of Anius Modestus

VII. 13. 10, 11 ***Shop***

VII. 13. 13

LATRINE

(Fiorelli p115)

VII. 13. 8, 14 ***House of L. Caecilius Communis***

LATRINE by entrance #14 in the rear of a large well appointed house.

(Eschebach p334: Fiorelli 115)

VII. 13. 15 ***Cella meretricia***

VII. 13. 16 ***Cella meretricia***

VII. 13, 19-21 ***Cornerhouse with brothel***

DOWN PIPE in outer wall

NB Arch construction probably over cess pit

Eschebach latrine in 20

VII. 13. 22 ***Shop***

VII. 13. 23 ***Shop***

Latrine (Eschebach p336)

DOWN PIPE

VII. 13. 24 ***Café of Suettius***

LATRINE

DOWN PIPE

(Eschebach p337)

VII. 13. 25, 1

DOWN PIPE

VII. 13. 3, 4, 16-18

VII. 13. 5

VII. 13. 6

VII.13.7

VII.13.11 Downpipe in outer wall

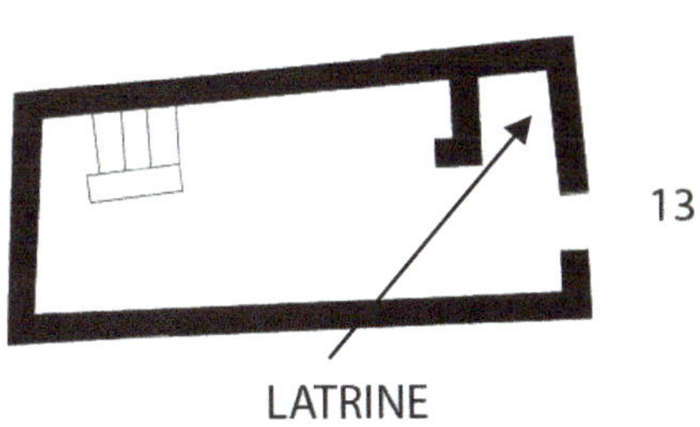

VII. 13. 13

VII.13.14 Latrine

VII. 13. 14 Water Pipe emerging through wall and groove of kerbstone

VII. 13. 14 Lead pipe going into wall from south to deliver water to latrine

VII. 13. 14 Hole in southern wall of latrine room for lead pipe to deliver water

VII. 13. 14 Groove in kerbstone for lead pipe coming from street and going through wall into house

VII.13.14 Downpipe

VII. 13. 19 Down pipe

VII. 13. 24 Down pipe

VII.13.23 Down pipe

VII. 13. 24

VII. 13. 25 Down pipe

REGIO VII 14

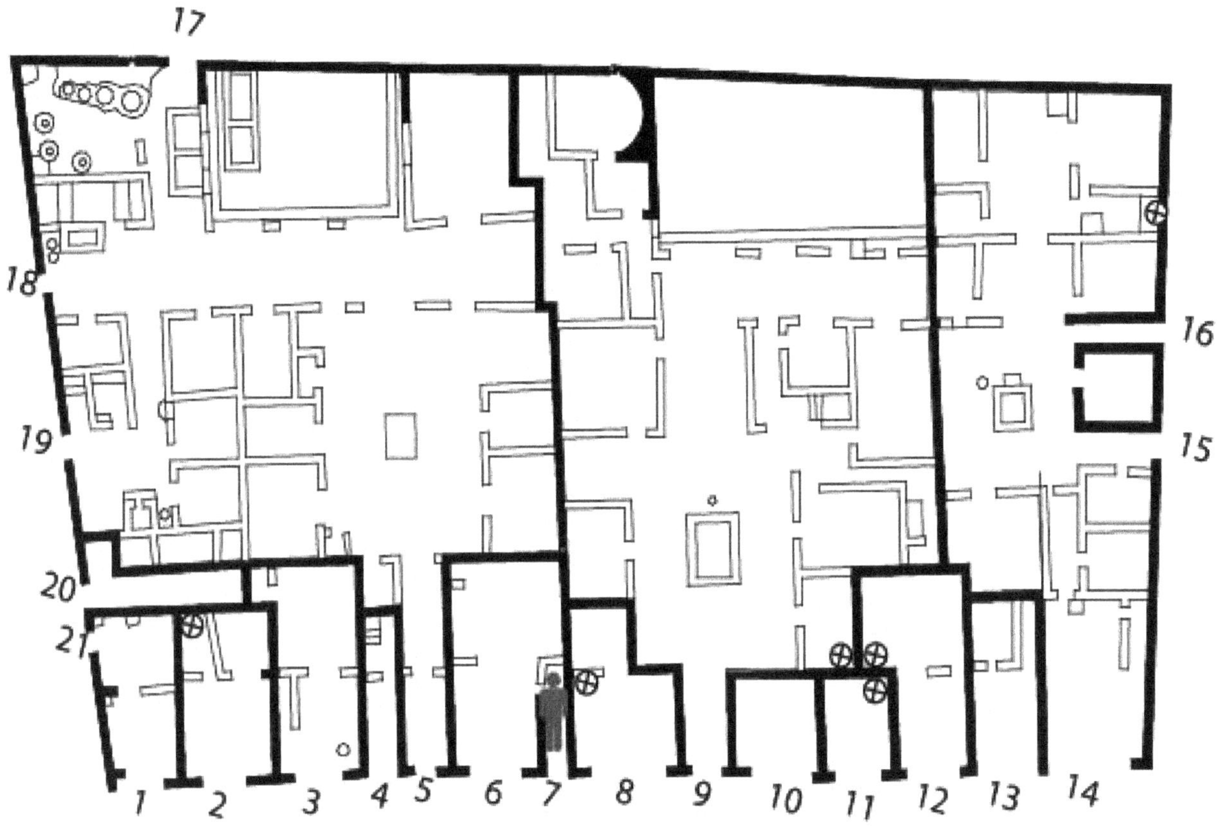

There were difficulties in obtaining access to some of the properties in this insula due to building works associated with VII.14.14-16. Also it was impossible to gain access to VII.14.5,17-19. Eschebach says that there are twelve latrines in this insula, including one 'public latrine' (VII.14.4)

This survey, though incomplete, has not found most of these latrines

There is considerable evidence for upper storey latrines in second storey accommodation particularly lining the Via Abbondanza

VII. 14. 1, 20 *Shop*

Latrine not confirmed

(Eschebach p337: Fiorelli 116)

Nothing

VII. 14. 2 *Shop*

Latrine not confirmed

(Eschebach p337: Fiorelli 116)

DOWN PIPE

VII. 14. 3 *Shop*

Latrine not confirmed

(Eschebach p338: Fiorelli 116)

Small room at rear but it has a solid floor

VII. 14. 4 Stairs with PUBLIC LATRINE beneath

No evidence (Eschebach p338)

VII. 14. 5, 17-19 *Casa del Cambio.*

NO ACCESS

Latrine in 19 (Eschebach p338: Fiorelli p117)

VII. 14. 6, 7 *Workshop*

Possible in long narrow room but no evidence

Latrine in 6 not confirmed

(Eschebach p339: Fiorelli p117)

VII. 14. 8 *Shop*

DOWN PIPE

VII. 14. 9 *House of V Popidius*

Latrine (Eschebach p339)

Not confirmed

DOWN PIPE

VII. 14. 10 *Shop*

Nothing

VII. 14. 11 *Shop*

?Latrine (Eschebach p340)

No latrine but a DOWNPIPE

VII. 14. 12 *Shop*

Latrine (Eschebach p340)

Not confirmed

DOWN PIPE

VII. 14. 13 *Shop*

POSSIBLE in long narrow room

(Eschebach p340: Fiorelli p117)

VII. 14. 14-16 *House of Caecilius Communis*

LATRINE & TWO DOWN PIPES in room 16

Latrine in 14 (Eschebach p340: Fiorelli p117)

VII.14.1 Downpipe

VII. 14. 2 Down Pipe

VII.14.3 Downpipe

VI.14.4 Latrine

VII. 14. 8 Down Pipe

VII. 14. 9 Down Pipe

VII. 14. 11 Down Pipe

VII. 14. 12 Down pipe

VII. 14. 13

VII.14.14 Latrine & 2 Down pipes

REGIO VII 15

An interesting insula. Two definite latrines and a number of down pipes of different sizes

VII. 15. 1, 2, 15 ***Double house***

Latrine in 2 (Eschebach p341: Fiorelli p118)

VII. 15. 3 ***House of the art traders***

LATRINE

(Eschebach p343: Fiorelli p117)

VII. 15. 4, 5

Workshop with café

Two latrines (Eschebach p343: Fiorelli p119)

LATRINE in 4

DOWN PIPE with PROBABLE

LATRINE in 5

VII. 15. 6, 7 ***Workshop with tavern***

Nothing

DOWN PIPE in outer wall

VII. 15. 8 ***Workshop of Heracla and Aegle***

Nothing

VII. 15. 9, 10 *Workshop with bar*

Latrine in 9 (Eschebach p344)

VII. 15. 11 ***House of Verus***

Latrine (Eschebach p344: Fiorelli p120)

VII. 15. 12-14 ***Double houses of A Octavius Primus and C Julius Nicephorus***

Latrine in 13 (Eschebach p345)

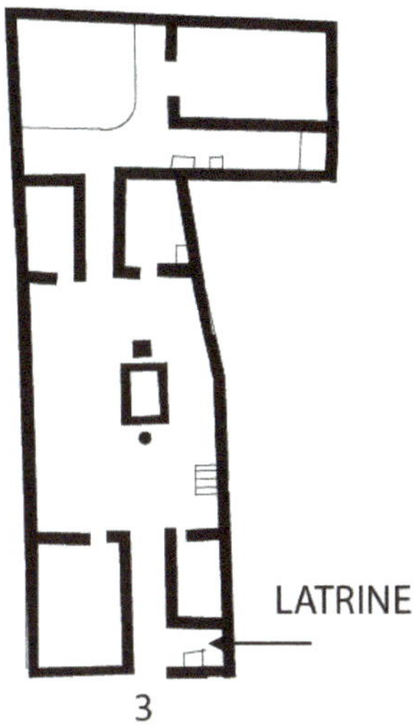

VII. 15. 3 Latrine

VII. 15. 2, 15 Possible down pipe

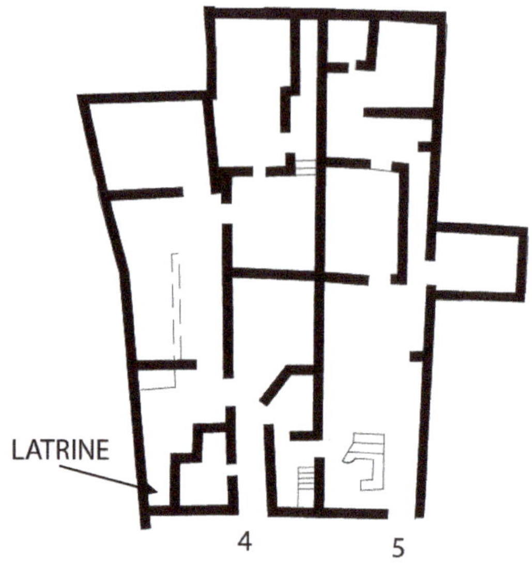

VII. 15. 4 Latrine

VII. 15. 6/7 Narrow down pipe in outer wall between #6 & #7

VII. 15. 5 Down pipe with possible latrine at rear of house

VII. 15. 9, 10 Down pipe

VII. 15. 10, 11

VII. 15. 10, 11

VII. 15. 11, 12 Downpipe in outer wall

REGIO VII.16 & 17 (Insula Occidentalis)

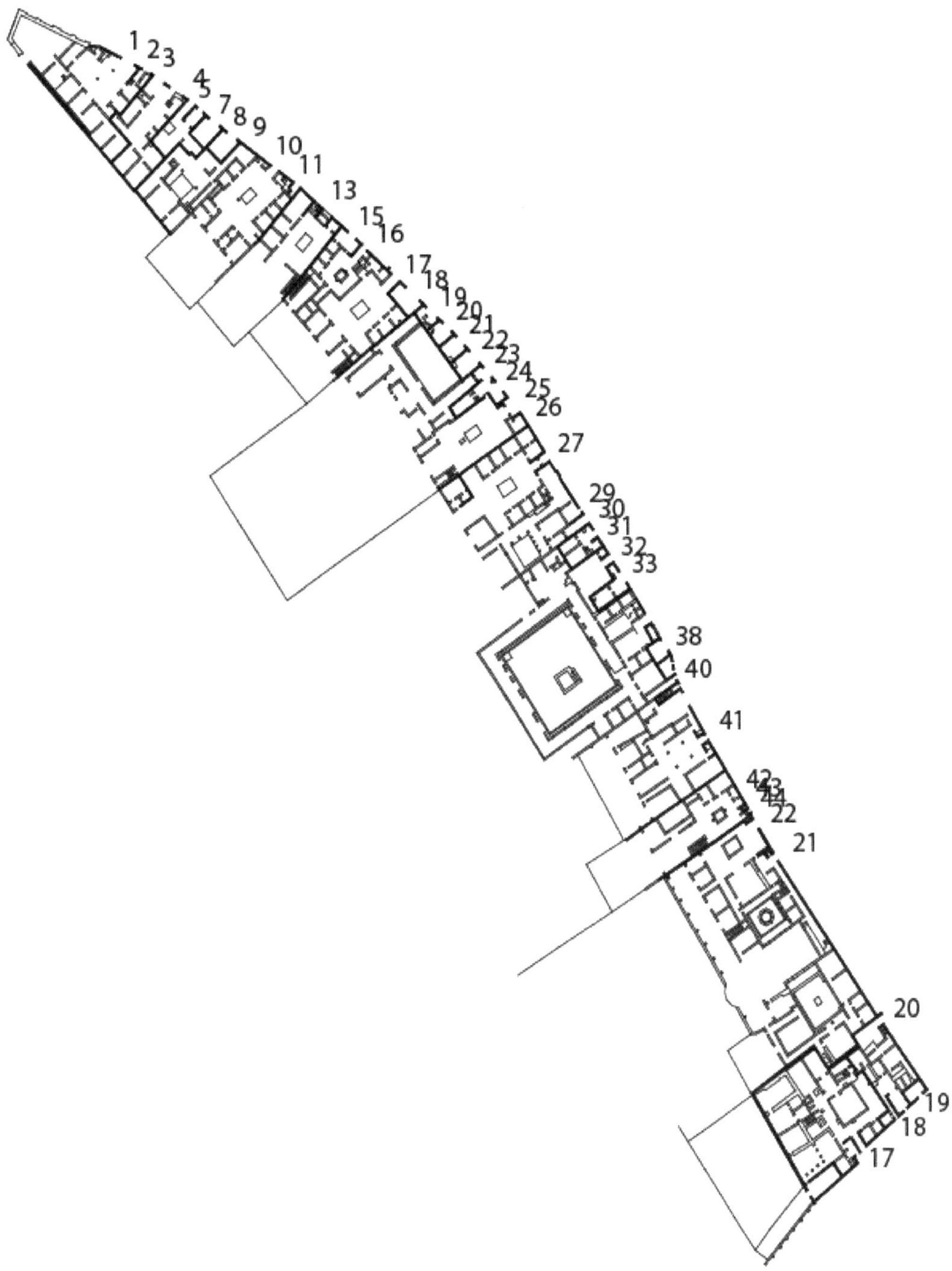

Only one down pipe (in VII.16.11)

VII.16.1Large House

VII.16.2 Shop

VII.16.3,4 House of Aemilius Crescens

VII.16.5,6 Mill house

VII.16.7,8 Wine tavern

VII.16.9 Small pistrinum

VII.16.10.11Work area with shop
DOWN PIPE in outer wall

VII.16.12-16

Apartments of Umbricius Scaurus

Latrine in 12 and 14 (Eschebach p348)

VII.16.17,20-22 Apartments of Fabius Rufus

VII.16.18 Small workshop

VII.16.11 Down pipe

REGIO VIII

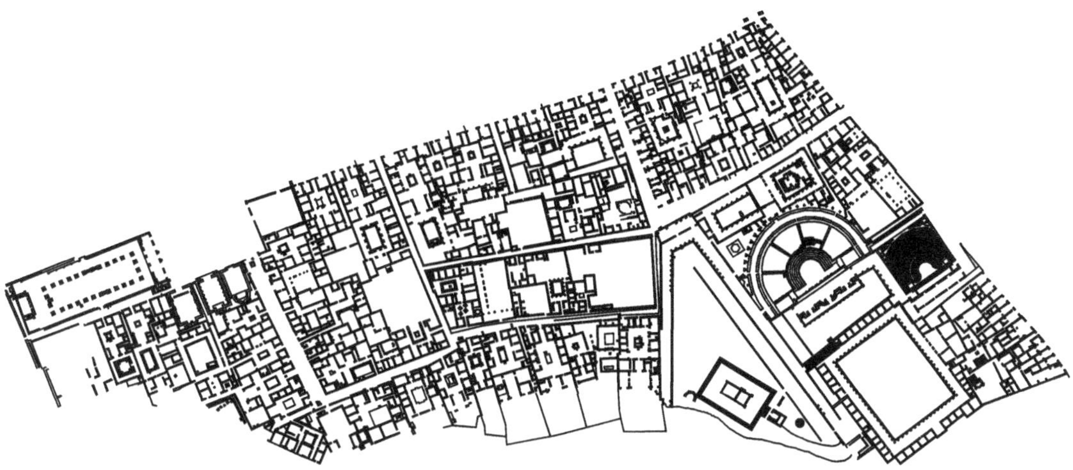

There are no latrines in VIII.1

REGIO VIII. 2

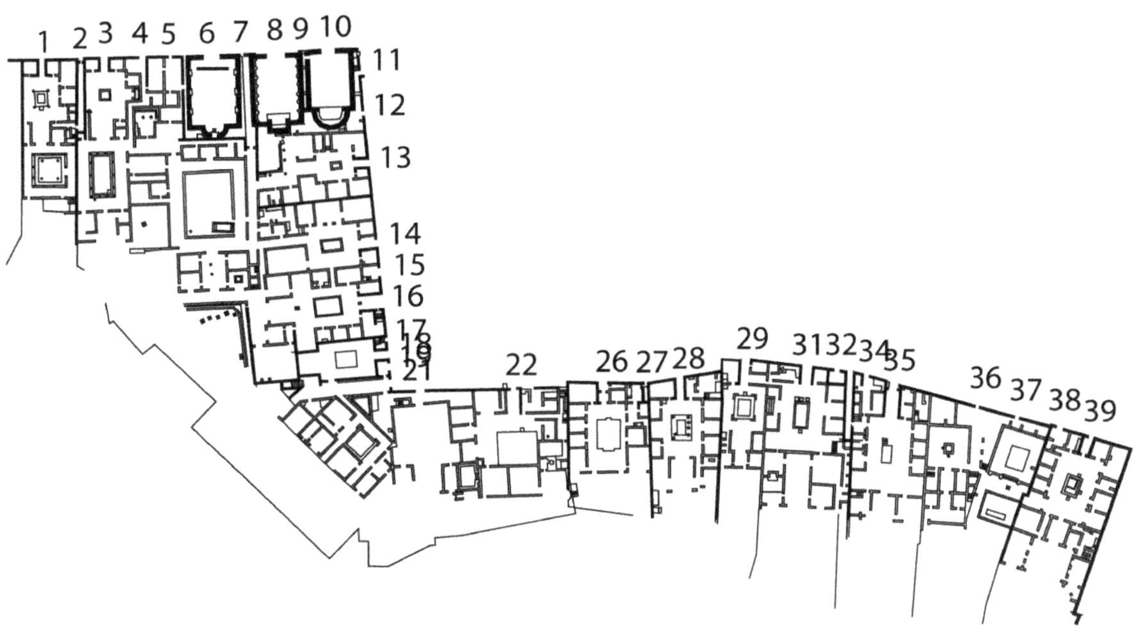

No latrines found. Eschebach lists three. There are two down pipes in VIII.2.38

VIII. 2. 1, 2 ***Casa di Championnet I***

Nothing

VIII. 2. 3-5 ***Casa di Championnet II***

Nothing

VIII. 2. 6-12 ***Curia***

Nothing

VIII. 2. 13 ***Dwelling House***

Nothing

VIII. 2. 14-20 ***Building Complex of L. Aelius Magnus***

Nothing

VIII. 2. 15 ***Shop***

Nothing

VIII. 2. 21 ***House of Aelius Magnus***

Nothing

VIII. 2. 22-24 ***Palaestra and Men's Bath***

Nothing

Latrine in 23 (Eschebach p358)

VIII. 2. 25 ***Altar***

VIII. 2. 26, 27 ***House of ? and the Vespinus***

Nothing

VIII. 2. 28 ***House with Nymphaeum***

Latrine in kitchen (Eschebach p359)

VIII. 2. 29, 30 ***House of Severus***

Nothing

VIII. 2. 31 Shop

Nothing

VIII. 2. 32-34 ***House of the Mosaic Column***

Latrine under stairs (Eschebach p360)

VIII. 2. 35 ***Shop***

Nothing

VIII. 2. 36, 37 ***House of Caecilius Phoebus***

Nothing

VIII. 2. 38, 39 ***House of Family Tullia***

TWO DOWNPIPES on east wall

Latrine in kitchen of 39 (Eschebach p361)

VIII.2.38 Two downpipes

REGIO VIII. 3

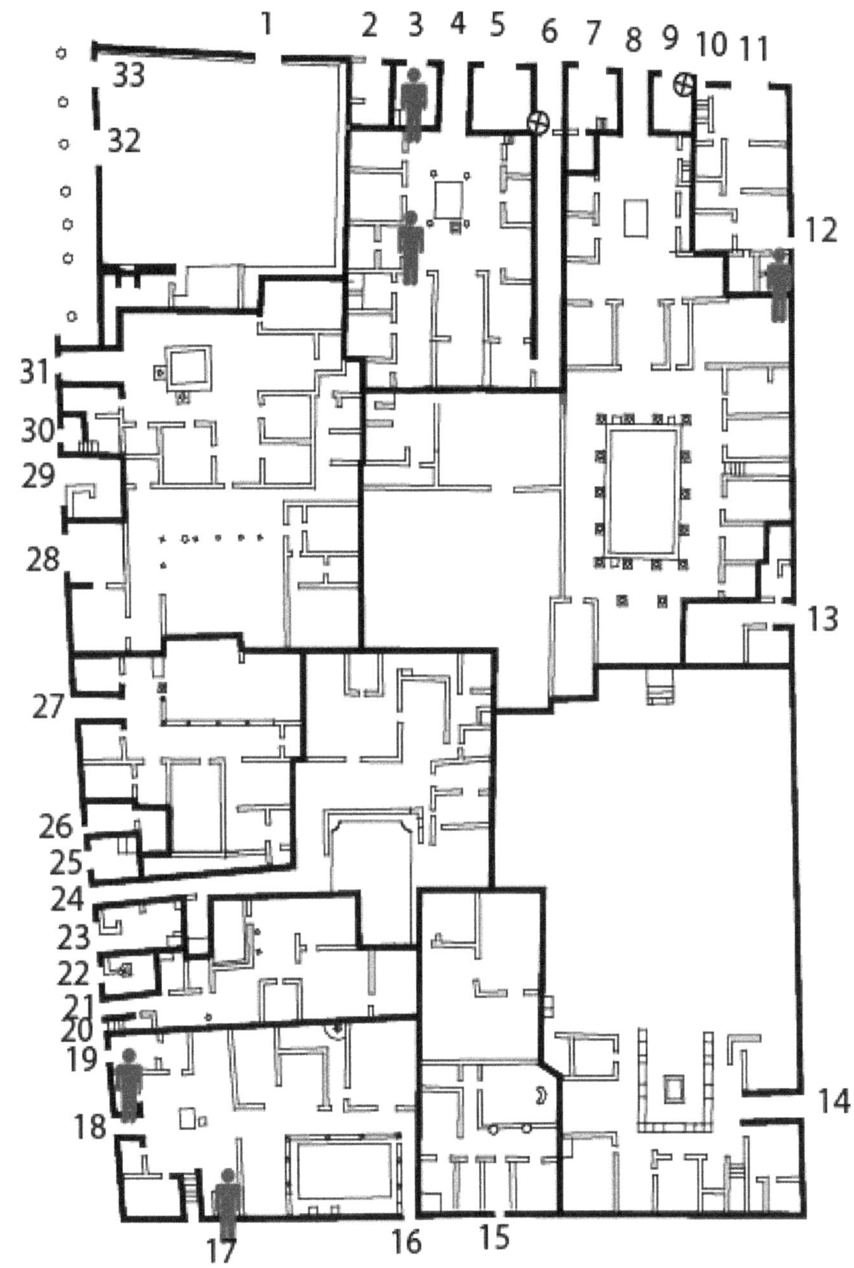

Five latrines in this insula, one of which is underground and three down pipes

VIII. 3. 1, 32, 33 *Comitium*

Nothing

VIII. 3. 2 *Shop*

Nothing

VIII. 3. 3 *Shop*

LATRINE

VIII. 3. 4, 6 *House of the Popidii*

LATRINE in cellar

VIII. 3. 5 *Shop*

Nothing

VIII. 3. 7 *Shop*

Nothing

VIII. 3. 8, 13 *House of Cinghiale*

Nothing

Latrine (Eschebach p364)

VIII. 3. 9 *Shop of Anteros*

DOWN PIPE in wall by entrance

VIII. 3. 10, 11, 12

LATRINE

Latrine in 12 (Eschebach p364)

VIII. 3. 13 *Small apartment*

Possible to right of entrance #13

Latrine (Eschebach p365)

VIII. 3. 14, 15 *House of Adonis*

LATRINE in 14

(Eschebach p365:Fiorelli p124)

VIII. 3. 16, 17, 18 *House of Diana*

NICHE LATRINE in 16

Probable pipe in the masonry construction to the right of the niche

NICHE LATRINE in kitchen of 17 with drainage from the latrine running down a pipe in the wall of the 'cellar' below

Latrine in 18 (Eschebach p366)

DOWN PIPE in outer wall

VIII. 3. 19 *Shop*

Nothing

VIII. 3. 20, 21 *House of Sex. Decimius Rufus*

NO ACCESS

VIII. 3. 22 *Shop*

Nothing

VIII. 3. 23 *Shop*

Latrine (Eschebach p367))

Possible latrine

VIII. 3. 24 *House of Apolline and Coronide*

DOWN PIPE in 24 in kitchen

Latrine (Eschebach p367:Fiorelli p125)

VIII. 3. 25 *Shop*

Nothing

VIII. 3. 26 *Small workshop*

Nothing

VIII. 3. 27 *Private house*

Nothing

VIII. 3. 28, 30, 31 *Shop and House of Pan*

Nothing

VIII. 3. 29 *Thermopolium*

Latrine (Eschebach p368)

VIII. 3. 3

VIII.3.4 Latrine in cellar

VIII. 3. 9

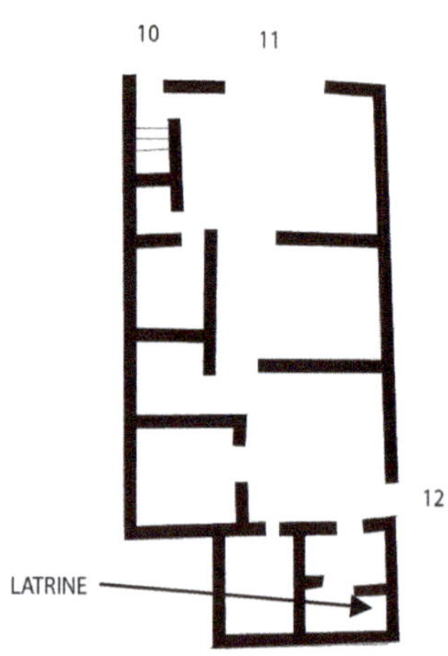

VIII. 3. 12

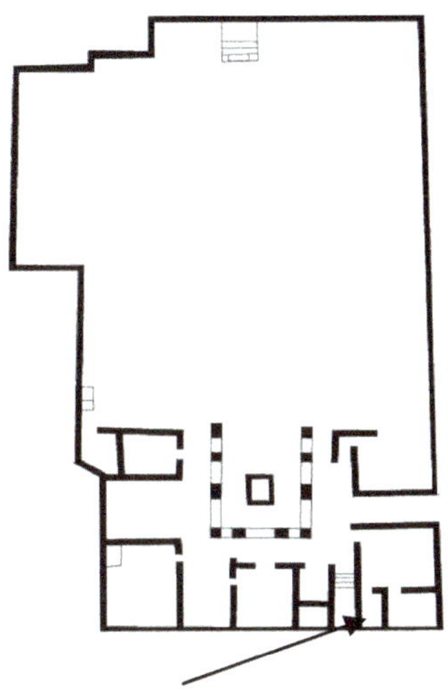

VIII. 3. 14

VIII. 3. 16

VIII. 3. 17

VIII. 3. 17 Downpipe through cellar wall

VIII. 3. 21,24

VIII. 3. 23 Possible latrine

REGIO VIII.4

An insula with shops lining two sides, four large houses and some workshops including a bakery.

VIII. 4. 1, 53 ***Tavern with workshop***
Latrine (Eschebach p369: Fiorelli p126)
(Also See Dickmann and Pirson 2004 p305)
Nothing found

VIII. 4. 2-6, 49, 50 ***House of QQ, Postumii, Proculus & Modestus***
2 Nothing: 3 Nothing
4 Latrine (Eschebach p369: Fiorelli p128)
The latrine in room 22 (not found) drained into the sewer from the Stabian Baths (Dickmann & Pirson 2004 p305)
5 Nothing
6 Nothing

VIII. 4. 7 ***Shop of Severus***
LATRINE with large diameter DOWN PIPE

VIII. 4. 8, 9 ***House of T. Mescinius Gelo***
Latrines in 8 & 9 (Eschebach p371: Fiorelli p128)

VIII. 4. 10 ***Shop***
DOWN PIPE in room b

VIII.4.11 ***Workshop***
DOWN PIPE SCAR
Latrine (Eschebach p371)

VIII. 4. 12 ***'Ganeum' and Brothel***
DOWN PIPE at top of short staircase

VIII. 4. 13 ***'Tonstrina'***
Latrine (Eschebach p372)
DOWN PIPE

VIII. 4. 14-16, 22, 23, 30 ***House of the Cornelii***
Latrine in 22 (Eschebach p372)

VIII. 4. 17, 18 ***Small workplace with two shops***
Nothing

VIII. 4. 19 ***Popina***

VIII. 4. 20 ***Workshop and shop***
Latrine (Eschebach p375)

VIII. 4. 21 ***Shop***

VIII. 4. 22, 23 ***Sacellum***

VIII. 4. 25 ***Shop***

VIII. 4. 26-29 ***House with Pistrinum of Felix with two shops***

VIII. 4. 30 ***Shop***

VIII. 4. 31-33 ***House with bar and shop***

VIII. 4. 34 ***Casa di Omfale***
LATRINE and DOWN PIPE (Eschebach p376)

VIII. 4. 35 ***Shop***

VIII. 4. 36 ***Large shop***
Latrine (Eschebach p376)

VIII. 4. 37 ***House in the row***
LATRINE & TWO DOWN PIPES & UPPER STOREY LATRINE
(Eschebach p376: Fiorelli p130)
(Also See Dickmann & Pirson 2004 p305)

VIII. 4. 38 ***Shop***

VIII. 4. 39 ***Small workshop and shop***

VIII. 4. 40, 40A ***Large shop***

DOWNPIPE at floor level in north east corner

(Latrine Dickmann & Pirson 2004 p305)

VIII. 4. 41 ***Roadside Toilet***

(Eschebach p377: Fiorelli p130)

VIII. 4. 42

VIII. 4. 43, 44 ***Shop***

VIII. 4. 45 ***Wine bar***

Latrine (Eschebach p378: Fiorelli p130: Dickmann & Pirson 2005 p305)

LATRINE in room d (No Photograph)

VIII. 4. 46

VIII. 4. 47 Large shop

Latrine (Eschebach p378: Fiorelli p130)

VIII. 4. 48

LATRINE

VIII. 4. 49.50

VIII. 4. 51, 52 ***Small Workplace with bar***

VIII. 4. 53

Possible rubbish disposal chute

VIII. 4. 7

VIII.4.7 Down pipe

VIII. 4. 10 Down pipe

VIII. 4. 11 Down pipe scar

VIII. 4. 12 Down pipe at top of stairs

VIII. 4. 13 Down pipe

Two down pipes & Latrine

VIII. 4. 34

VIII. 4. 37 Upper storey & down pipe

VIII.4.40 Pipe in ground

VIII.4.48 Latrine

VIII. 4. 45

VIII. 4. 53 Feature from above

VIII. 4. 53 ?for rubbish disposal

REGIO VIII. 5

An Insula with eighteen properties but mostly shops. Some of the insula is storage for plants for display. The position of the down pipes along the Via Abbondanza may suggest underground drainage

VIII. 5. 1. 40 *Taverna*

Nothing

VIII. 5. 2, 3, 5 *Casa del Gallo*

NO ACCESS

Latrine in 2 (Eschebach p379)

Latrine in 5 (Eschebach p380)

DOWN PIPE in SE corner of 3

VIII. 5. 4 *Shop*

DOWN PIPE in SE corner

VIII. 5. 6 *Shop*

Down pipe scar

VIII. 5. 7 *Large shop*

Nothing

VIII. 5. 8, 9, 11-14 *Atriumhouse*

LATRINE in 9 (Eschebach p381)

DOWN PIPE

VIII. 5. 10 *Shop*

DOWN PIPE

VIII. 5. 15, 16, 38 *Garden restaurant*

Latrine in 15,16 (Eschebach p 381)

DOWN PIPE in 15

Latrine in 38 (Eschebach p382)

VIII. 5. 17, 20 *Small workshop and bar*

DOWNPIPE

VIII. 5. 18, 19

Latrine in 18 (Eschebach p382)

VIII. 5. 21 *Shop*

DOWN PIPE

Latrine (Eschebach p382)

VIII. 5. 22, 23

Latrine in 22 (Eschebach p382)

DOWN PIPE

VIII. 5. 24-26 *House of the doctor I*

VIII. 5. 27 *Shop*

DOWN PIPE

VIII. 5. 28, 29, 34, 35 *House of Calce*

NO ACCESS

VIII. 5. 30, 31 *Shop*

DOWN PIPE

VIII. 5. 32 *Shop*

VIII. 5. 33 *Shop*

LATRINE

DOWN PIPE

VIII. 5. 34, 35 *Shop*

VIII. 5. 36 *Republican Baths*

VIII. 5. 37 *House of the family Fabia*

VIII. 5. 39 *House of Acceptus*

VIII. 5. 3 Down Pipe

VIII. 5. 4 Down pipe

VIII. 5. 6 Wall scar

VIII. 5. 9 Down pipe

VIII. 5. 10 Down pipe

VIII.5.14 Latrine

VIII. 5. 15

VIII.5.20 Downpipe

VIII. 5. 21 Down pipe

VIII. 5. 22 Down pipe

VIII. 5. 23 Down pipe

VIII. 5. 27 Room 1

VIII. 5. 30

VIII. 5. 33 room 1

VIII. 5. 39 Latrine in room e

REGIO VIII.6

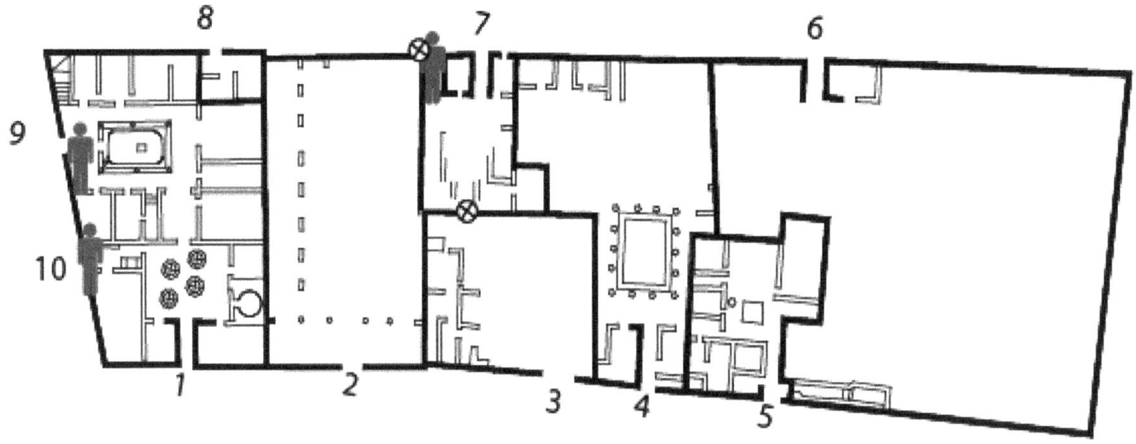

In this Insula there are only eight properties. Two entrances are to gardens.

VIII. 6. 1, 9-11 ***Pistrinum of P Emilius Gallicus***

LATRINE in 9

LATRINE in 10

VIII. 6. 2 ***Garden with buildings***

Nothing

VIII. 6. 3 ***Garden***

Latrine ?? cess pit (Eschebach p387)

Nothing

VIII. 6. 4 ***House with atrium***

Nothing

VIII. 6. 5 ***Living house***

Nothing

VIII. 6. 6 ***Garden***

Latrine (Eschebach p387)

Nothing

VIII. 6. 7 ***'Demolition house'***

LATRINE in room e

With UPPER STOREY LATRINE and DOWN PIPE

Notable because of square hole in plaster on rear wall filled with charcoal, probably a wooden post upon which to hang a garland

DOWN PIPE into cistern on southern wall of room b

VIII. 6. 1, 9-10

VIII.6.7 Water pipe into cistern

VIII. 6. 6 Latrine with down pipe

VIII.6.6 Upper storey latrine

VIII.6.11 Latrine

REGIO IX

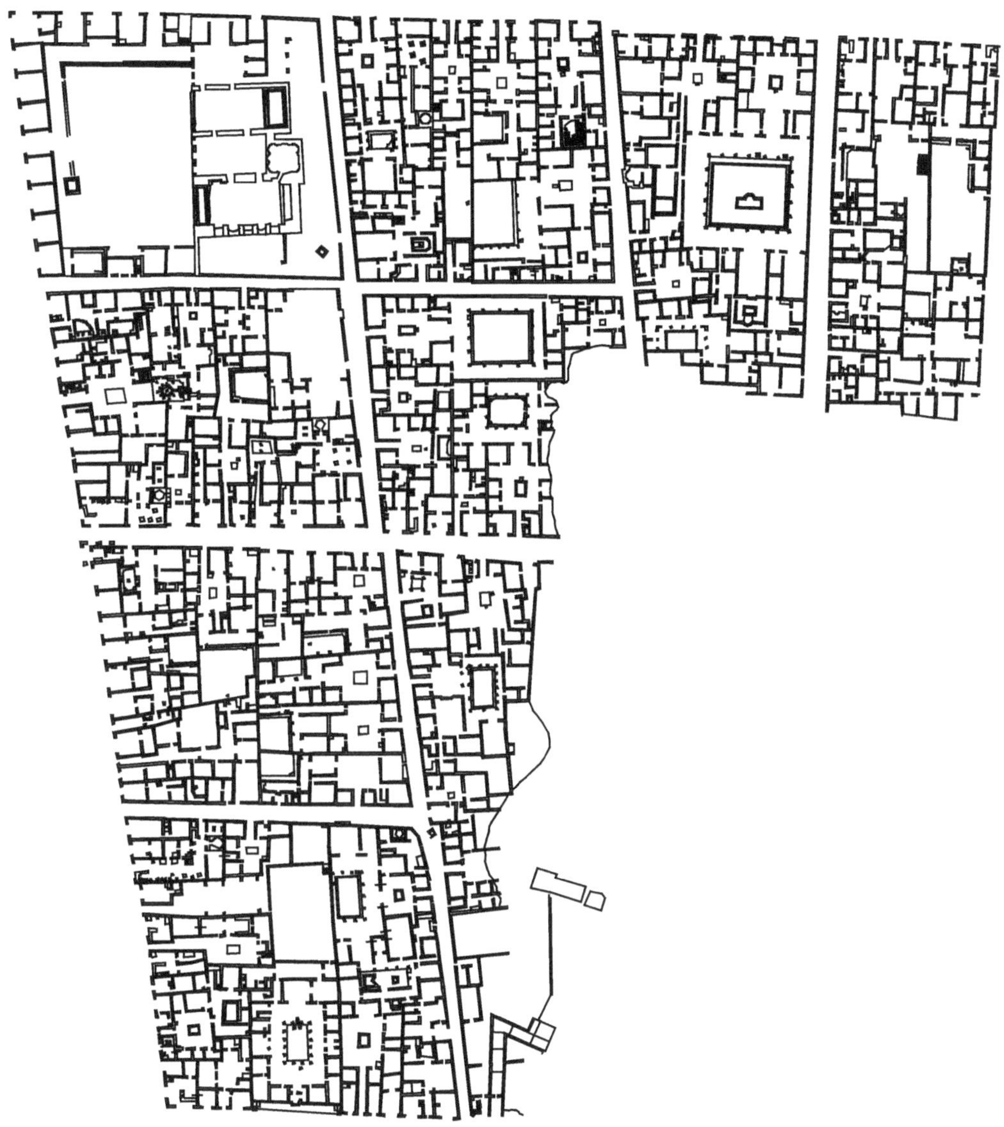

REGIO IX.1

A most interesting Regio with a number of upper storey niche latrines as well as down pipes and ground floor toilets

IX. 1. 1,34 *Small workshop with shop*

Latrine under stairs (Eschebach p397: Fiorelli p137)

No evidence

IX. 1. 2 *Shop*

Nothing

Latrine under stairs (Eschebach p397)

IX. 1. 3, 33 *Pistrinum*

Possible in 33 to right of entrance

Latrine (Eschebach p397: Fiorelli p137)

IX. 1. 4 *Shop*

LATRINE in kitchen (Eschebach p398) Extremely interesting construction with slots in walls apparently forming a triangular seat. No Privacy

IX. 1. 5 *Workplace of L Livius Firmus*

DOWN PIPE in outer wall

IX. 1. 6 *Thermopilium*

DOWN PIPE in north wall to left of entrance

Latrine under stairs (Eschebach p398: Fiorelli p137)

IX. 1. 7 *Workshop of Paccius Alexander,*

DOWN PIPE in southern wall

Latrine (Eschebach p399: Fiorelli p137)

IX. 1. 8 *Popina*

Nothing

Latrine (Eschebach p399: Fiorelli p137)

IX. 1. 9 *Shop of Fabritius, Metalworker*

DOUBLE DOWN PIPE inside wall to north of entrance

IX. 1. 10, 11 *Shop of Fabritius*

Nothing

Latrine in limestone niche in 11 (Eschebach p399)

IX. 1. 12 *Atriumhouse*

Nothing

IX. 1. 13 *Thermopolium*

Nothing

Larine (Eschebach p 400)

IX. 1. 14 *Workshop and shop*

Scar on north wall

IX. 1. 15, 16 *Thermopolium of Primus*

Nothing

IX. 1. 17, 18, 19 *Workplace of Lollia Successa*

DOWN PIPE at rear of property, room f

Latrine under stairs in 18 (Eschebach p401)

IX. 1. 20, 30 *House of MM Epidii Rufus and Sabinus*

LATRINE: probably two seater, in kitchen (Eschebach p401)

IX. 1. 21 *Shop*

Possible scar on back wall

Latrine (Eschebach p402)

IX. 1. 22, 29 *Double house of CC. Cuspii, Pansa and Proculus*

LATRINE in Room t. Plastered wall

LATRINE in room h

Two latrines in 22 (Eschebach p 402)

Latrine in 29 (Eschebach p402)

IX. 1. 23 *Shop*

DOWN PIPE on west wall

Latrine (Eschebach p403)

IX. 1. 24 *Shop*

DOWN PIPE in north east corner

IX. 1. 25, 26 *House of the Jews*

Nothing

Latrine in25 (Eschebach p403)

IX. 1. 26 MODERN OFFICE

IX. 1. 27 *Shop of Pacuvius Erasistratus*

Nothing

IX. 1. 28 *Stabulum of Thesmus*

DOWN PIPE in outer east wall.

NB Obviously discharges onto road.

Note the worn groove in the kerb stone

IX. 1. 31, 32 *Small bar*

NICHE LATRINE in room to left of entrance (Eschebach p404: Fiorelli p140)

DOWN PIPE into kitchen work surface

IX. 1. 34

DOWN PIPE in wall to east of #34

IX. 1. 1 Room a

IX.1.3,33 Latrine under staircase

IX. 1. 4

IX. 1. 5

IX. 1. 6

IX. 1. 9

IX. 1. 20 Room x

IX.1.17,18,19

IX. 1. 20, 30 Two seater latrine in room x

IX. 1. 22 Room h

IX. 1. 22 Room t

IX.1.23

IX.1.24

IX. 1. 28

IX. 1. 31

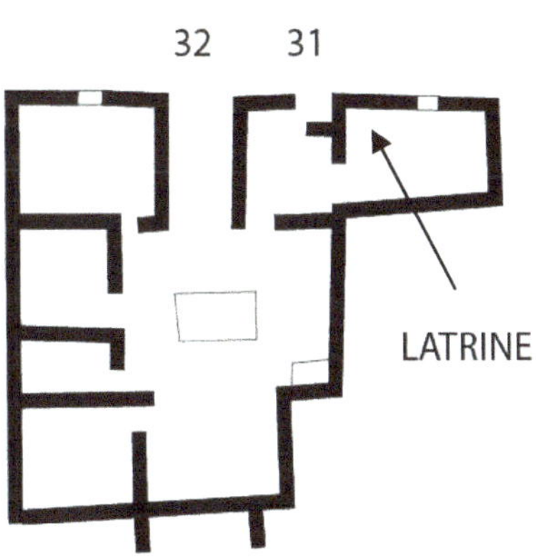

IX.1.31 Pipe on kitchen surface

Down pipe in street wall by #34

REGIO IX.2

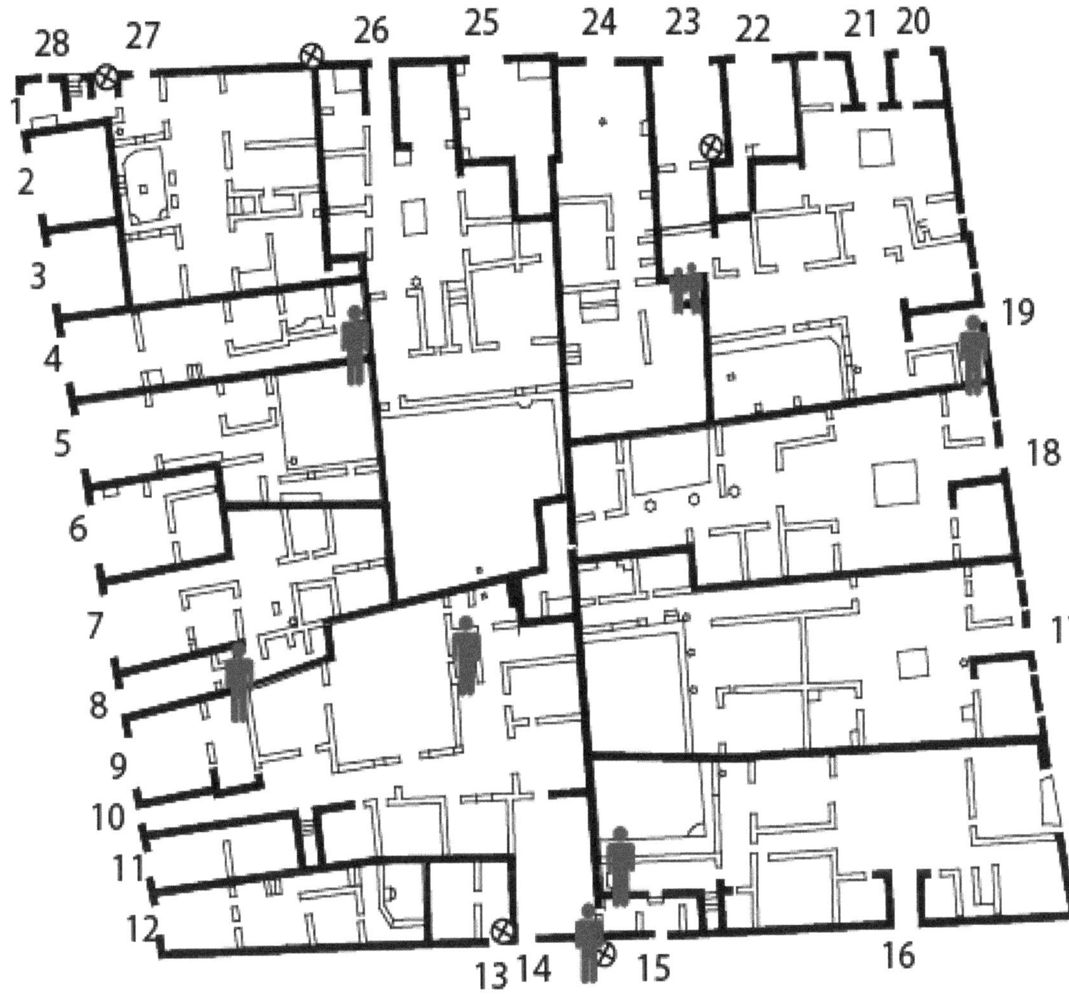

An insula with twenty properties. Three upper storey latrines visible: IX.2.8, IX.2.15 and IX.2.19,21. The balcony latrine at IX.2.15 is of special interest

IX. 2. 1 ***Colonnade***

Nothing

IX. 2. 2 ***Shop with workshop***

Nothing

IX. 2. 3 ***Shop***

Nothing

IX. 2. 4 ***Shop with workshop***

LATRINE (Eschebach p405)

IX. 2. 5 ***Shop with workshop***

Nothing Latrine in 5 (Fiorelli p141)

IX. 2. 6 ***Shop of Hilario***

Nothing

IX. 2. 7, 8 ***Casa della Fontana Amore***

UPPER STOREY LATRINE in 8

Latrine (Fiorelli p141)

IX. 2. 9 ***Shop***

Nothing

IX. 2. 10, 14 ***House of Chlorus***

LATRINE; may be two seater

(Eschebach p407)

IX. 2. 11 ***Tavern of Attiorum I***

Nothing

IX. 2. 12 ***Tavern of Attiorum 2***

Nothing

Latrine (Eschebach p407: Fiorelli p142)

IX. 2. 13 ***Small workplace***

DOWN PIPE (? Rainwater) in south east corner.

DOWN PIPE in outer wall #15

IX. 2. 15, 16 ***House of Dentatius Panthera***

Latrine in 16 (Eschebach p408: Fiorelli p143)

LATRINE; probably two seater

Also LATRINE on balcony

IX. 2. 17 ***House of Q Bri(U)ttius Balbus***

Nothing

IX. 2. 18 ***House of Curvius Marcellus***

Possible in kitchen room o. Latrine (Eschebach p409)

IX. 2. 19-21 ***Atriumhouse with shop***

UPPER STOREY LATRINE

IX. 2. 22 ***Shop***

Nothing

IX. 2. 23 ***Shop***

Latrine (Eschebach p 411)

Nothing

IX. 2. 24 ***Stable***

LATRINE probably two seat.

(Fiorelli p144)

IX. 2. 25 ***Caupona of Thyrsus***

Nothing

IX. 2. 26 ***House of M Casellius Marcellus***

NO ACCESS.

DOWN PIPE in wall IX.2.26/27

IX. 2. 27, 28 ***House of the marriage of Neptune and Amphytrite***

Latrine in 28 (Eschebach p412: Fiorelli p144)

DOWN PIPE in 28

IX. 2. 1, 28, 29

IX. 2. Wall between 2&3 Two down pipes

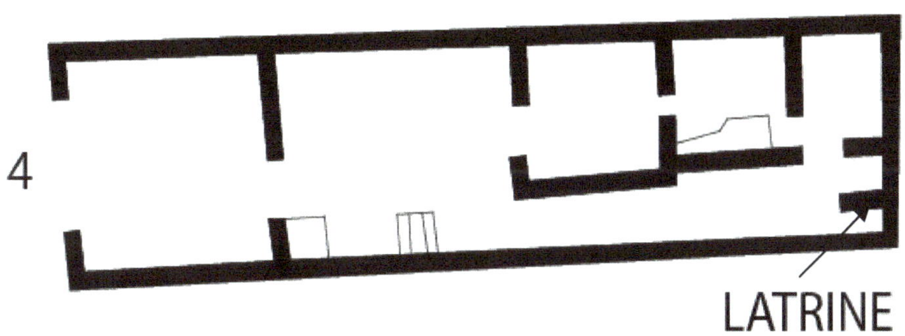

IX.2.4

IX. 2. 8 ?waste disposal

IX. 2. 8 Upper storey niche latrine

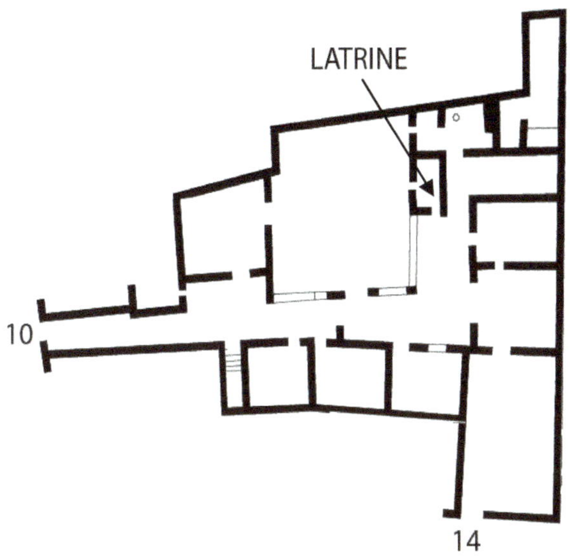

IX. 2. 10, 14

IX. 2. 13 Pipe

IX.2.15 Upper storey Latrine in outer wall (balcony) with down pipe

IX. 2. 15 Outer wall down pipe from balcony latrine

IX. 2. 15, 16

IX .2. 19-21 Upper storey Latrine

IX.2.23 Down pipe in buttress, room a

IX. 2. 24

IX. 2. 26/27 Pipe in outer wall

IX. 2. 28 Down pipe

IX.2.5 Pipe in wall of garden

REGIO IX.3

This plan may not be accurate for #24 and #25

several latrines and down pipes. Two of the latrines are in the second storey and the double down pipes suggest latrines in second and third storeys

IX. 3. 1, 2 *Office of Ubonius*

DOWN PIPE on outer wall of room 4

Latrine (Eschebach p412:Fiorelli p145)

IX. 3. 3 *Shop*

Nothing

IX. 3. 4 *Shop*

UPPER STOREY LATRINE with associated DOWN PIPE

IX. 3. 5, 24 *House of the Suonatrici*

DOWN PIPE

Latrine(Eschebach p413)

TWO LATRINES (No Photographs)

IX. 3. 6-9 ***Show-place for Pius IX***

Nothing

3. 6 ***Shop of doctor Phosphorus***

Nothing

3. 7 ***Shop of Philocalus***

Eschebach latrine

Nothing

3. 8 ***Shop of Titus***

Nothing

3. 9 ***Workshop of Titus***

Nothing

IX. 3. 10-12 ***Pistrinum of T. Terentius Proculus***

Latrine in 12 (Eschebach p414:Fiorelli p146)

IX. 3. 13 ***Caupona of Fabius Celer***

Nothing

Latrine (Eschebach p416:Fiorelli p146)

IX. 3. 14 Shop

DOWN PIPE Encased in masonry

LATRINE in kitchen

latrine in kitchen (Eschebach p416)

IX. 3. 15 ***House of Philocalus***

DOUBLE DOWN PIPE in room n

Latrine under stairs of 16 (Eschebach p416:Fiorelli p146)

DOWN PIPE (metal) in Fauces, room a

IX. 3. 16 ***Office of C. Tettius Faustus***

Nothing

IX. 3. 17 ***Small workshop of Q. Sallustius Inventus***

UPPER STOREY LATRINE with DOWN PIPE

Latrine (Eschebach p417:Fiorelli p146)

IX. 3. 18 ***Small workshop of P Paccius Clarus***

LATRINE in northeast corner

Latrine (Eschebach p417)

IX. 3. 19, 20 ***Pistrinum of*** *Pyramus*

Nothing

IX. 3. 21, 22, 23 ***Plot of land of A Vettius Caprasius Felix***

Latrine in 23(Eschebach p418:Fiorelli p147)

IX. 3. 25 ***House of L. Clodius Varus***

LATRINE

(Eschebach p419)

IX. 3. 4 Upper storey latrine with down pipe in wall

IX. 3. 4 ground floor level

IX. 3. 5 Down Pipe

IX. 3. 14 Down pipe encased in masonry

IX. 3. 14

IX. 3. 15

IX.3.15 Upper storey latrine

IX.3.15 Iron Down pipe

IX. 3. 17 Down pipe from upper storey latrine

LATRINE

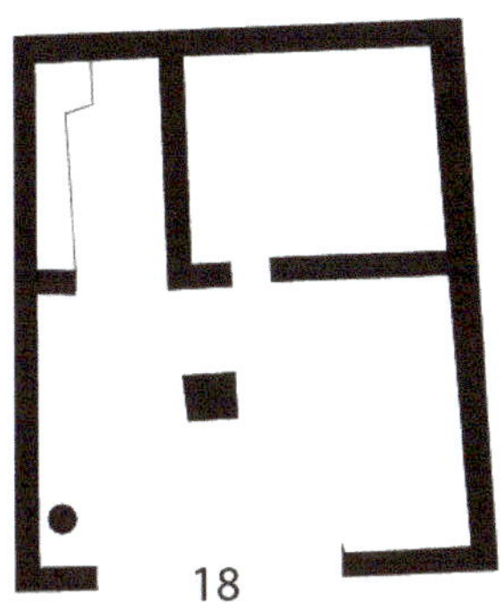

IX. 3. 18 Probable latrine

Note the down pipe encased in masonry

IX.3.25

REGIO IX.4

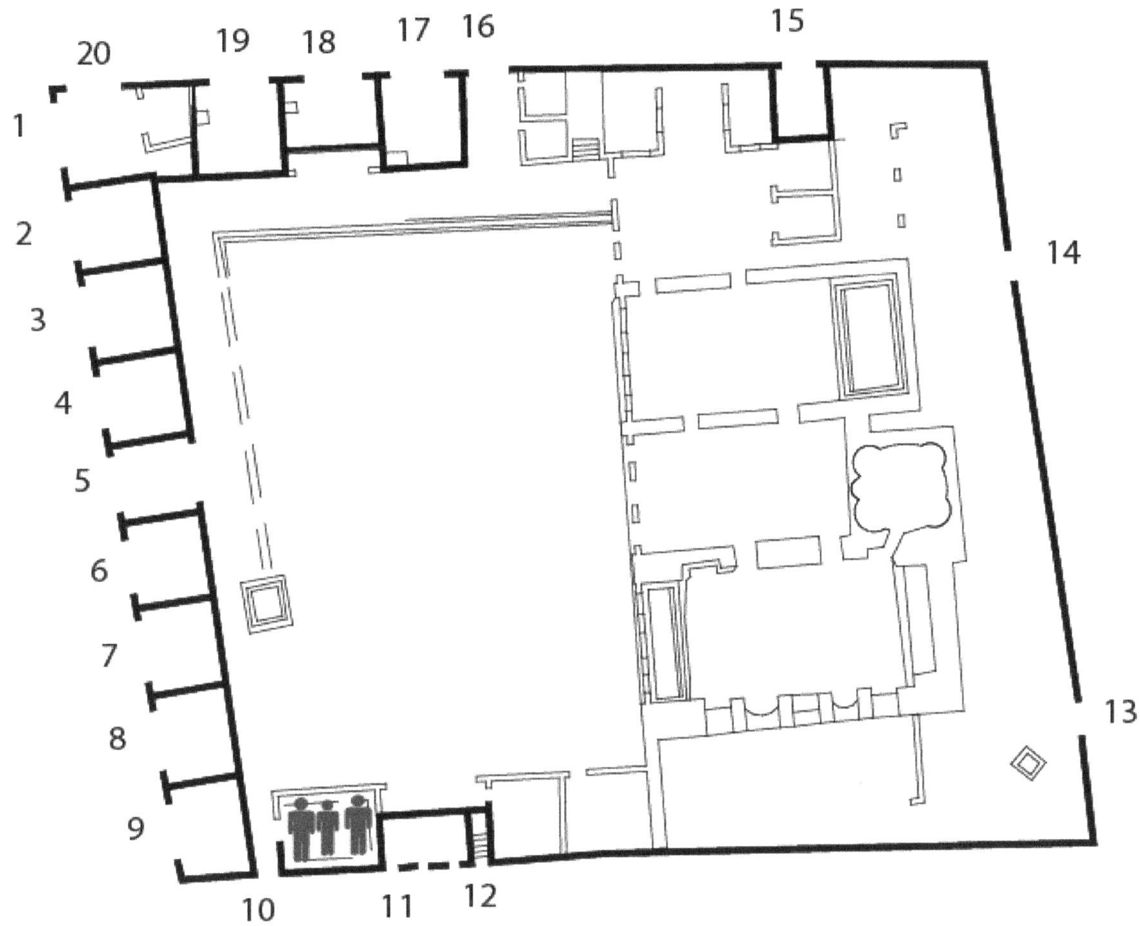

Apart from the multi seat latrine at IX.4.10 there is no evidence either of latrines or down pipes in the whole of this insula. Almost certainly this is because the insula was under reconstruction in 79CE

IX .4 .5,10,15,16,18 Central Baths

IX.4.10 Multiple seat latrine (Eschebach p420)

There is little evidence of this without excavation

IX.4.10 Multi-seat latrine in Central Baths

REGIO IX.5

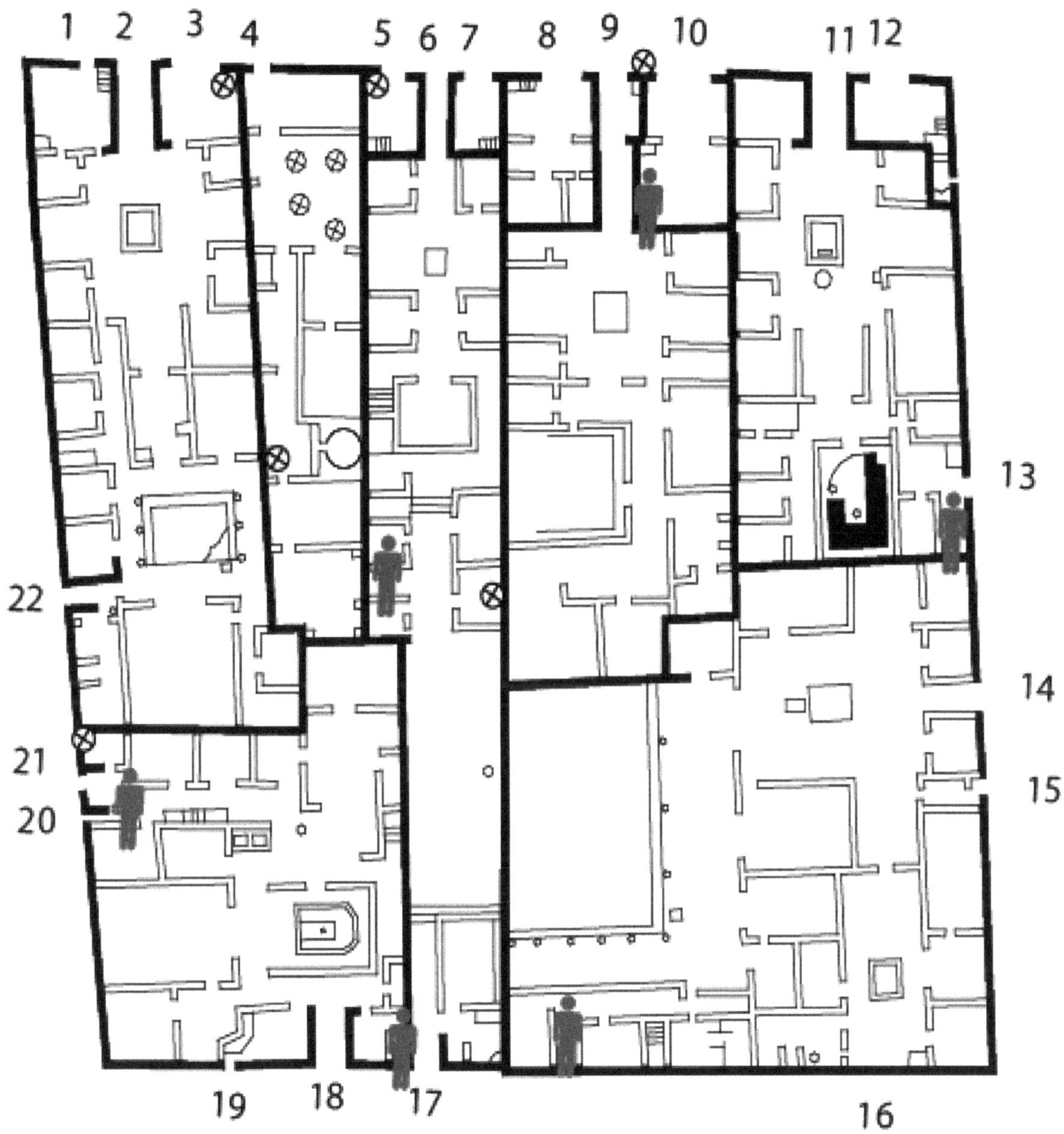

IX.5.1,2,3,22 ***House of the Skeleton (House of Achilles)***

Nothing in 1

DOWN PIPE and possible LATRINE in 3.

2,22

Latrine in 2 (Eschebach p422)

IX.5.4 ***Pistrinum***

WATER PIPE into storage for bakery

Latrine (Eschebach p422)

IX.5.5 ***Small shop***

DOWN PIPE in north west corner

IX.5.6,17 ***House of the Wholesale Trader***

Latrine in 6 (Eschebach p423)

LATRINE

IX.5.7 ***Large shop***

DOWN PIPE in outside wall between 7 and 8

IX.5.8 ***Shop***

Nothing

Latrine (Eschebach p423)

IX.5.9 ***House of Poppaeus Sabinus (House of the Pygmies)***

Latrine in kitchen (Eschebach p424)

Possible

IX.5.10 ***Shop***

LATRINE (Eschebach p422)

IX.5.11,13 ***House of Poppaeus Primus***

LATRINE seen through door 13 (Eschebach p424)

IX.5.12 ***Shop***

Nothing

IX.5.14-16 ***House of the restaurant and brothel***

LATRINE in 14 room 8 (Eschebach p425)

DOWN PIPE in 15 Room 1

IX.5.17 ***Stable behind #6***

IX.5.18-21 ***House of Jason***

LATRINE with DOWN PIPE in eastern wall (Eschebach p425)

IX.5.20 ***Entrance to Stairs***

Probable latrine under stairs

IX.5.21/22

DOWN PIPE in outer wall

UPPER STOREY NICHE LATRINE

Despite almost no remaining features this is almost certainly an upper storey latrine

IX.5.3 Down pipe

IX.5.4 Water pipe into storage for bakery

IX.5.5 Down Pipe

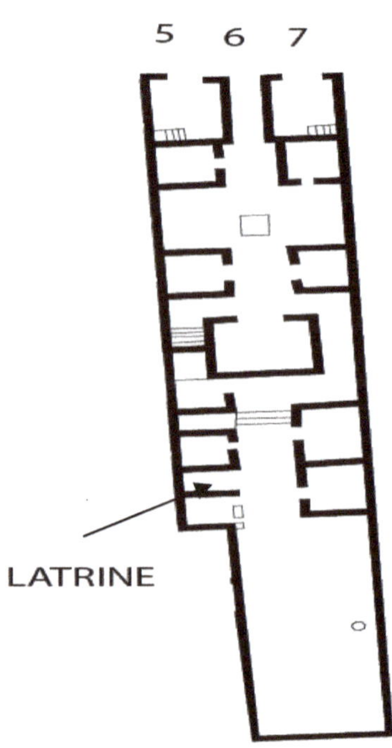

IX.5.6

IX.5.9 Possible latrine

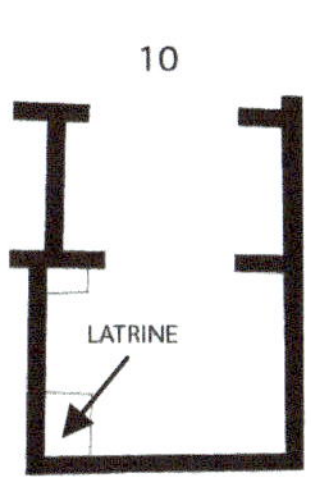

IX.5.10

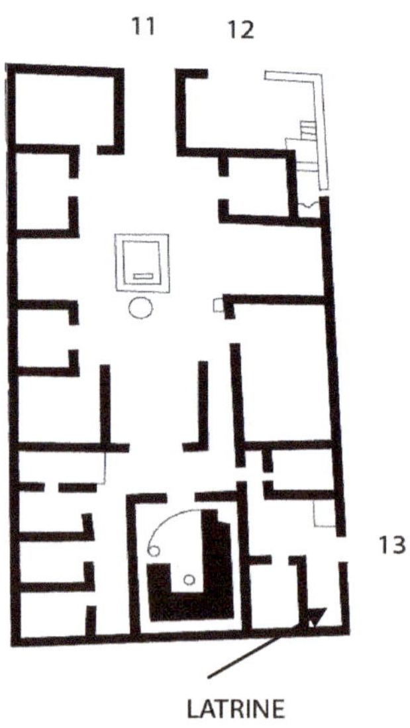

IX. 5. 13

IX.5.14 room 8

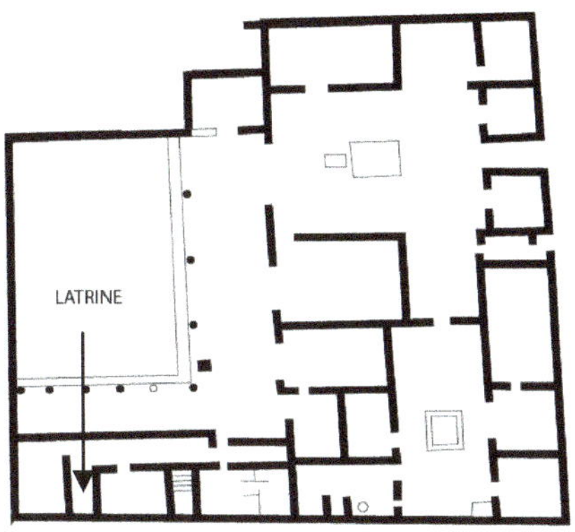

IX.5.14,15,16

IX.5.18 Latrine with pipe in wall

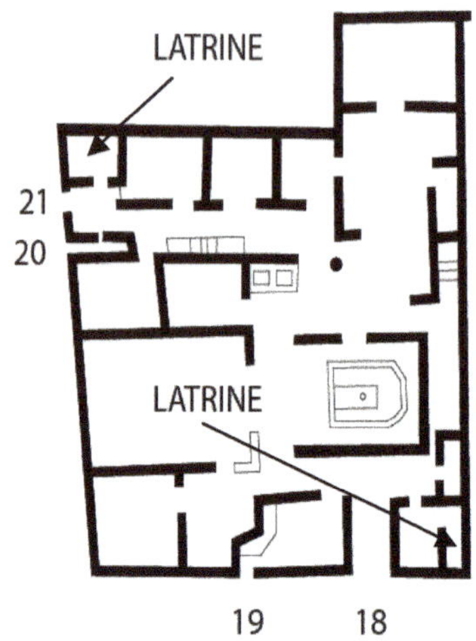

IX.5.21 Upper storey latrine

IX.5.20 Probable latrine

IX.5.21/22 Down Pipe in outer wall

REGIO IX.6

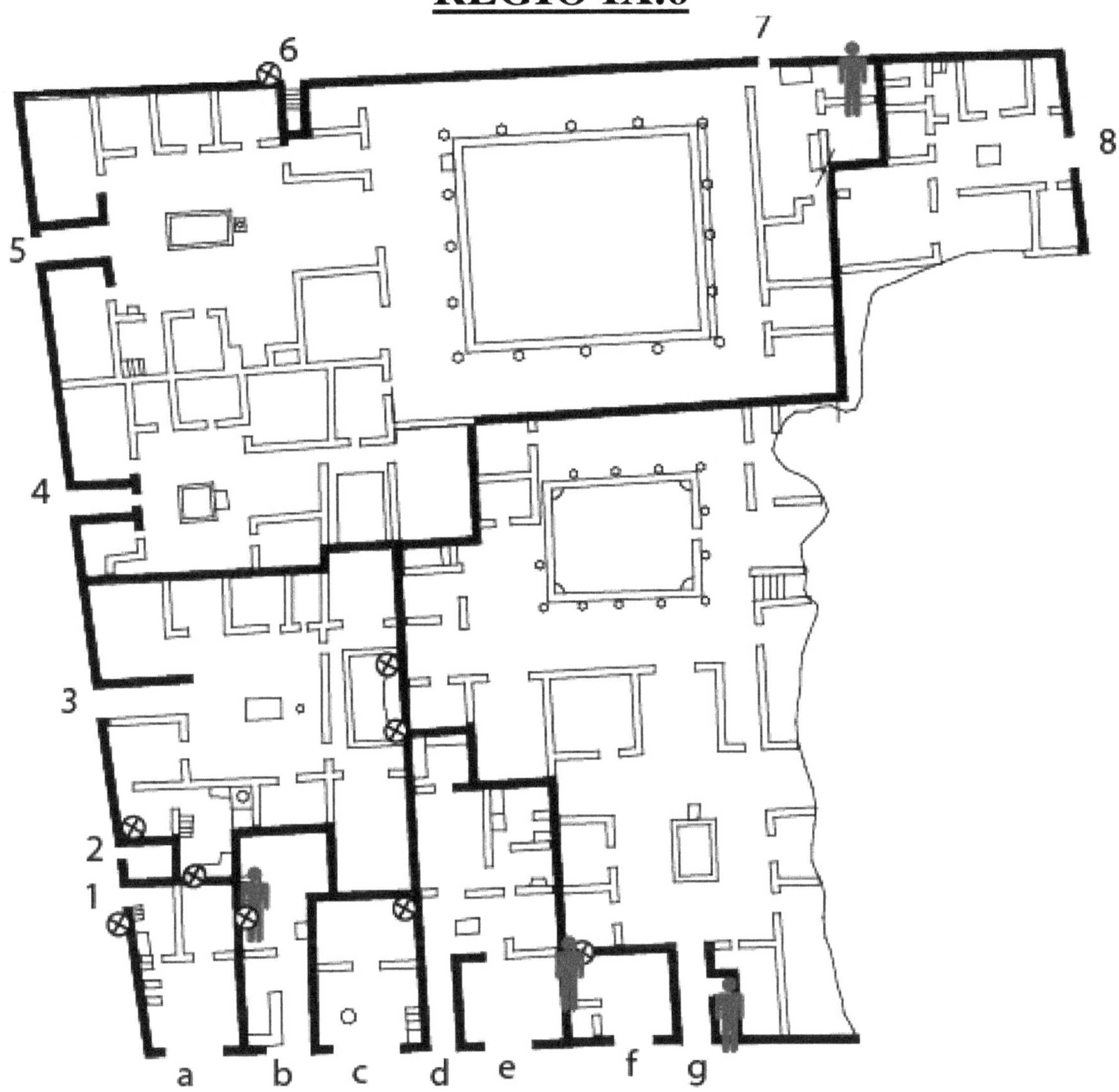

The numbering of this insula is confusing with both numbers and letters appearing on the Eschebach plan.

IX.6.1,a ***Shop with workshop of Marcus***

DOWN PIPE at top of wall by #1

Latrine (Eschebach p428)

IX.6.2 ***Cella Meretricia***

Nothing

IX.6.3 ***House of P. F. L.***

DOWN PIPE

Latrine (Eschebach p426)

IX.6.4-7 ***House of Jucundus & Quartilla***

Latrine in 7 (Eschebach p422)

Possible in room at north east of property

IX.6.6 ***Stairs***

DOWN PIPE in outer wall

IX.6.8 Brothel of Amandus

Nothing

IX.6.b ***Caupona of Marcus(Ex gladiator)***

Latrine (Eschebach p428)

LATRINE & UPPER STOREY NICHE & DOWN PIPE

IX.6.c Shop

Latrine (Eschebach p428)

DOWN PIPE

IX.6.d,e ***Casa di Dido & Aeneas***

Latrine (Eschebach p428)

Nothing

IX.6.f

LATRINE in NW corner (Eschebach p429)

IX.6.g

Possible

IX.6.1 Pipe at top of wall

IX.6.3 Down pipe

IX.6.3 Down pipe

IX.6.6 Down pipe

IX.6.1,a (wrong board # Not 6.6)

IX.6.b (wrong board # Not 6.7)

IX.6.c Down pipe

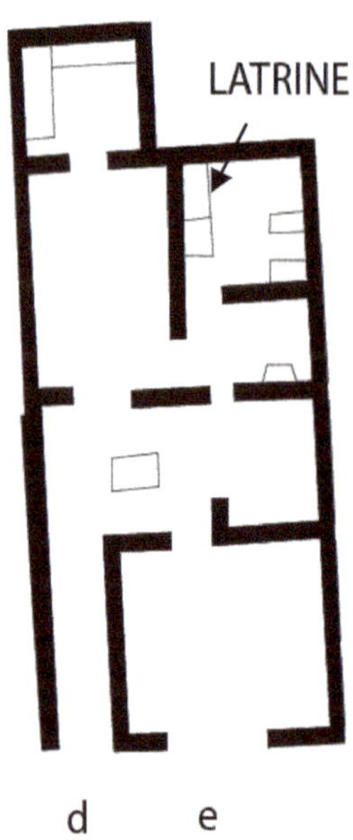

?IX.6.d,e

IX.6.f Latrine

IX.6.f Down pipe

REGIO IX.7

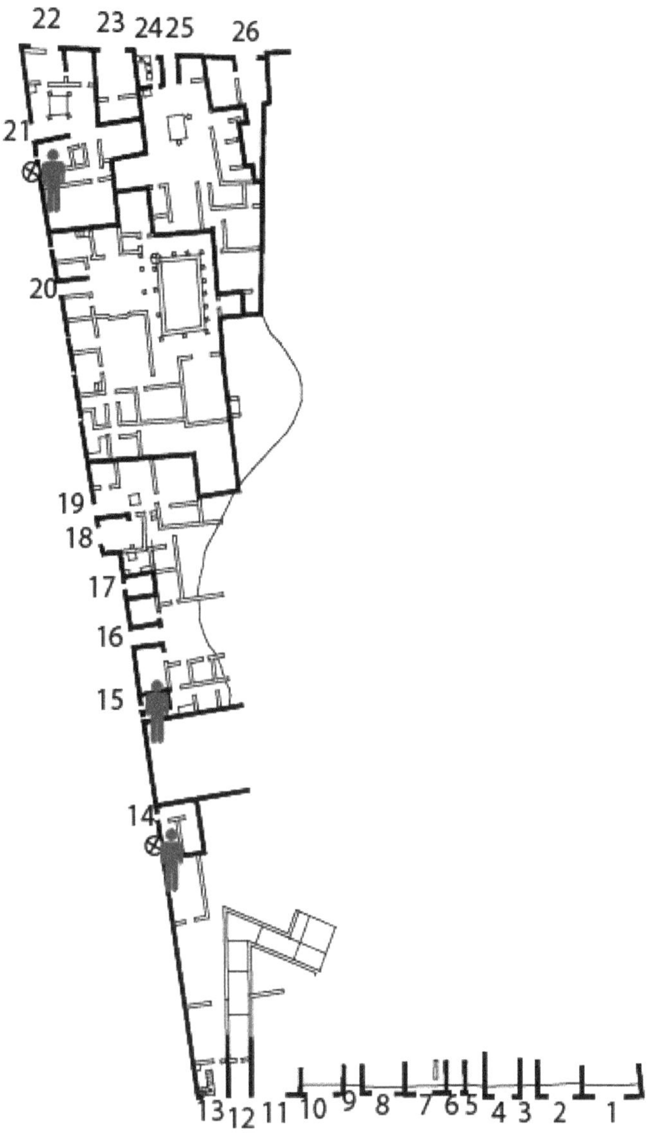

Numbers IX.7.1-11 are unexcavated and IX.7.12-19 only partially excavated.

Two upper storey latrines are visible in 14 and 15

IX.7.12 ***House of Polyphemus and Galathea***

Nothing

IX.7.13 ***Thermopolium***

Nothing

IX.7.14 ***?Back door***

UPPER STOREY LATRINE

DOWN PIPE in wall

IX.7.15 ***Cella meretricia***

Nothing

Latrine (Eschebach p433)

IX.7.16 House of A. Virnius Modestus

UPPER STOREY LATRINE

IX.7.17 ***Stairs Cella meretricia***

Latrine (Eschebach p433)

Nothing

IX.7.18,19 House of Specchio

Latrine (Eschebach p434)

Nothing

IX.7.20 ***House of Fortuna***

DOWN PIPE in outer wall

Latrine (Eschebach p434)

IX.7.21,22 ***Small hotel and bar of Tertius***

LATRINE (Eschebach p435)

IX.7.23 ***Bar of Ti. Claudius Epaphroditus***

Latrine (Eschebach p435)

Nothing

IX.7.24,25 ***Restaurant and bar of MM. Fabii Memor and Celer***

Nothing

IX.7.26 ***Lupanar of Fabius Memor***

Latrine (Eschebach p436)

Nothing

IX.7.14 Down Pipe

IX.7.14 Upper Storey Latrine

IX.7.16 Upper Storey Latrine and downpipe

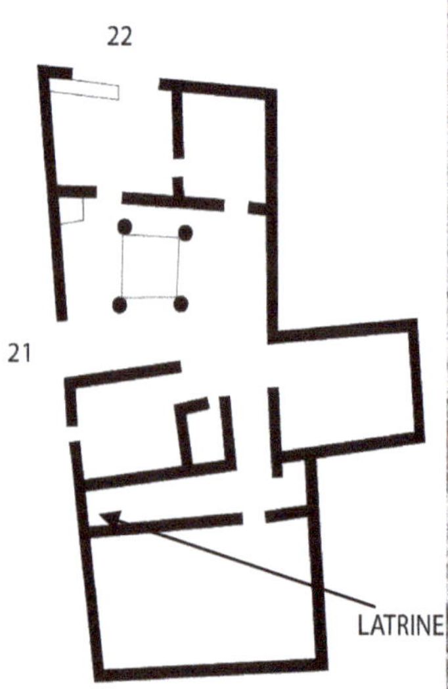

IX.7.21,22

IX.7.20 (21) Down pipe

REGIO IX.8

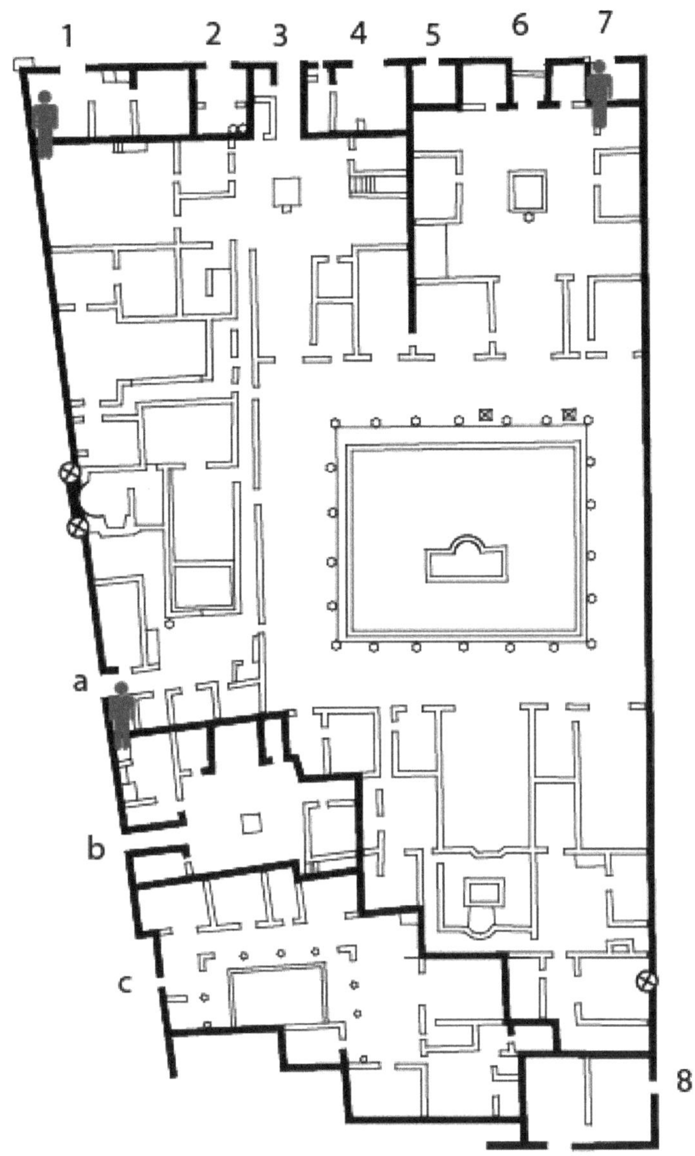

An insula which is mainly filled by the House of the Centenary but also has one other house and five small properties.

IX.8.1 ***Shop***

LATRINE

IX.8.2 ***Little garden of Potitus***

Latrine (Eschebach p437)

Nothing

IX.8.4 Shop

Latrine (Eschebach p437)

IX.8.3,6,a ***House of the Centenary***

LATRINE (Probable two seat)

(Eschebach p438)

IX.8.7 ***Small shop***

LATRINE in South west corner

IX.8.8 ***House of Primigredia***

Nothing

IX.8.b ***Inn of C Hyginius Firmus***

Possible in room to L of doorway #b

Latrine (Eschebach p439)

IX.8.c ***House***

Nothing

Two water DOWN PIPES for garden

NO PHOTOGRAPH

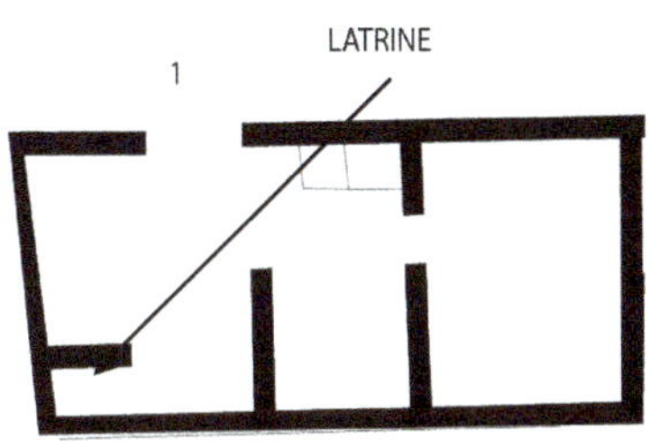

IX.8.1

IX.8.3,6,a House of the Centenary

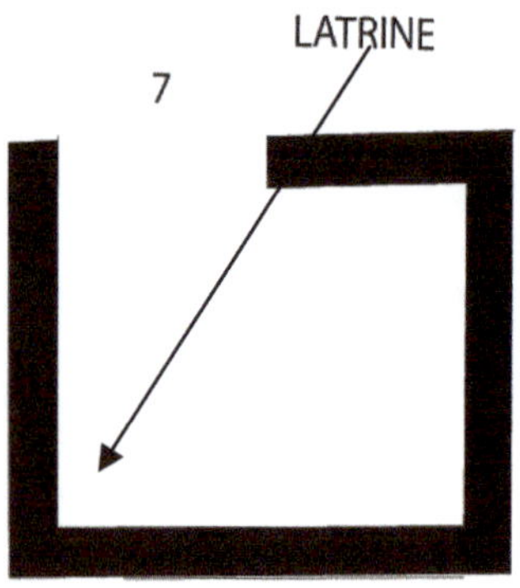

IX.8.7

REGIO IX.9

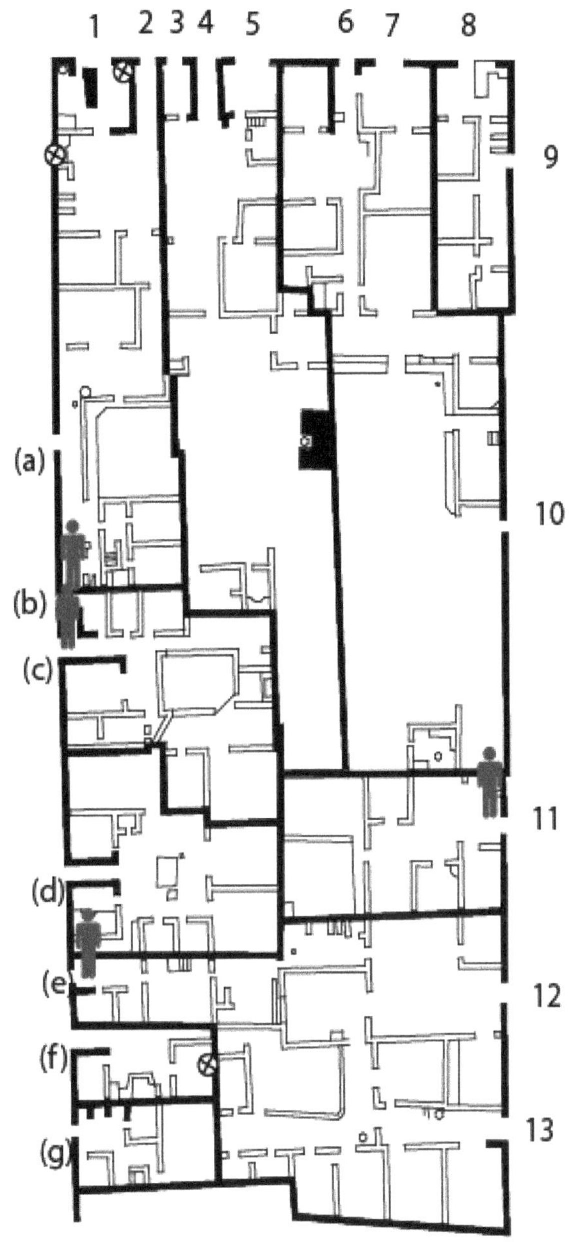

Ten properties. Three definite latrines.

NB 19=a: 18=c: 17=d: 16=e: 15=f & 14=g More confusion with board numbers

IX. 9. 1, 2, 19 ***Workshop with thermopolium and caupona of L Statius Receptus***

LATRINE (Eschebach p440)

NARROW BORE PIPE in room a

DOWN PIPE outer west wall room c

IX. 9. 3-5 ***House of the new doctor II***

Nothing in 3 or 5

4 NO ACCESS

Latrine (Eschebach p441)

IX. 9. 6, 7, 10 ***Workplace of the wine maker C. Caesius Restitutus***

Eschebach ?latrine in 6

Nothing

IX. 9. 8, 9 ***Workshop with caupona***

Possible against eastern wall

IX. 9. 11 ***Terraced house***

Latrine (Eschebach p442)

LATRINE with DOWN PIPE by pedestal

IX. 9. 12, 13 ***Stable***

Latrine in 13 (Eschebach p442)

Nothing

IX. 9. g (14) ***House of P. Aemelius Celer***

DOWN PIPE on top of wall between 13 & 14

IX. 9. f (15) ***The doctors house***

Latrine (Eschebach p443)

DOWN PIPE east wall room f

IX. 9. d (16) ***Living house***

LATRINE (Eschebach p443)

Nothing

IX. 9. b, c ***House of Sulpicius Rufus***

LATRINE to north of entrance (Eschebach p443)

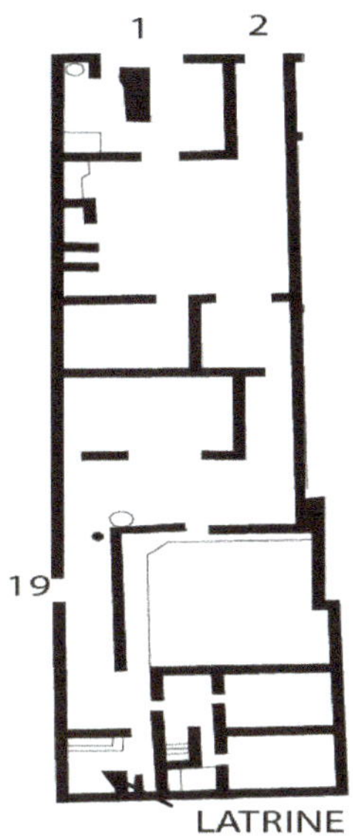

IX.9.1 Latrine

IX. 9.1 West wall downpipe

IX. 9.1 Narrow bore downpipe, room a

IX. 9.11Latrine

Down pipe on top of wall between 13 & 14

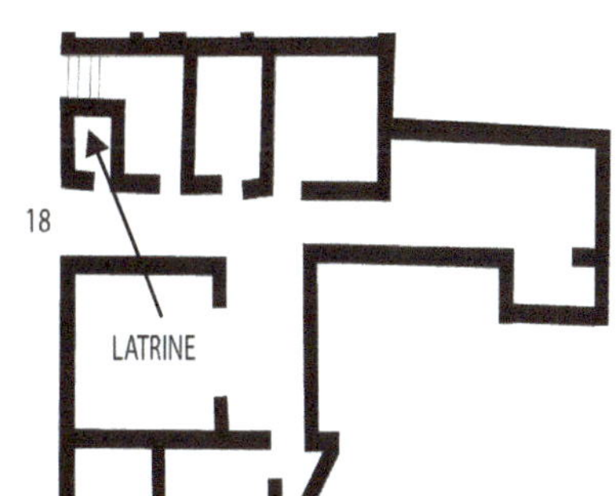

IX. 9. b,c (17,18)

IX.9.d(16) Latrine, room i

IX.9.f Downpipe in eastern wall

REGIO IX.13

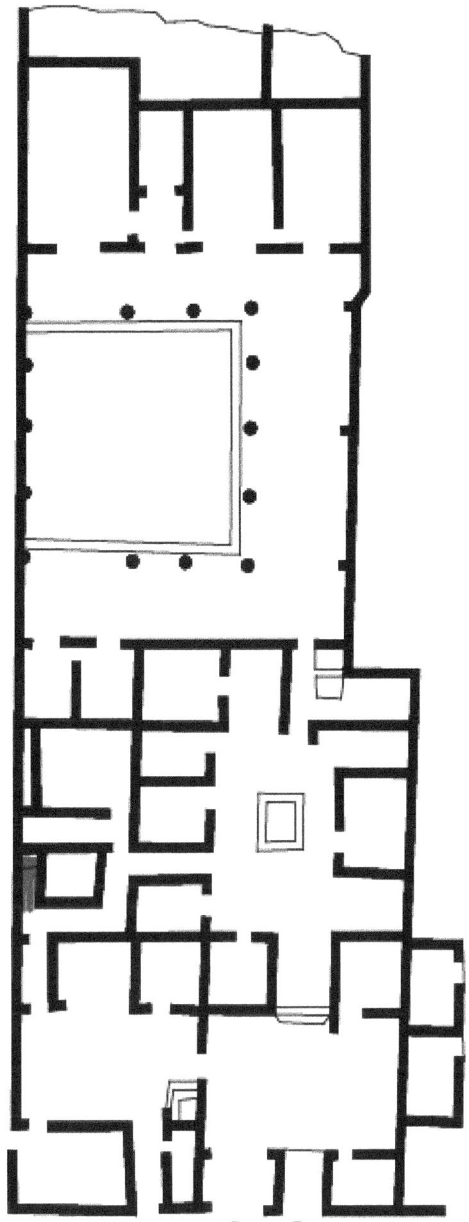

IX.13.1,2,3 UPPER STOREY LATRINE in the restored House of Julius Polybius

IX.13.1,2,3 Upper storey latrine

REFERENCES

Allison, P., Sear, F. 2002. *Casa dela Caccia Antica.* Band 11, Häuser in Pompeji. Deutsches Archäologisches Institut. Hirmar Verlag. München

Arthur, P. 1986. Problems of the urbanizaation of Pompeii: Excavations 1980-1981. *Antiquaries Journal* LXVI:I29-44.

Bartosiewicz, L. 2003. 'There's something rotten in the state:' Bad smells in Antiquity. European Journal of Archaeology Vol. 6 (2): 175-195

Bragantini, I. 1981. *Pitture e pavimenti di Pompei* Ministero per i beni culturali e ambientali, Istituto centrale per il catalogo e la documentazione. Roma

Breton, E. 1855. *Pompeia* Paris: Baudry

Clark, P., Davis, A. 1989. The power of dirt: an exploration of secular defilement in Anglo-Canadian culture *Canadian Review of Sociology and Anthropology* 26(4): 650-673.

Curtis, R. I. 1979. The garum shop of Pompeii. *Cronache Pompeiane Vol 5:5-23*

Dague, R. R. 1972. Fundamentals of odor control. *Journal of the Water pollution Control Federation* 48:583-94

Daremberg, C., Saglio, E. (eds.) 1873-1910. *Dictionaire des antiquites Grecques et Romaines.* Paris: Hachette.

De Haan, N. 2001. Pompeian private baths and the use of water. In: Koloski-Ostrow, A. O. (ed.) *Water use and hydraulics in the Roman city* AIA Colloquia and Conference Papers 3: 41-47. Boston.

De Jorio, A. 1820. Ricerche sul tempio di Serapide in Pozzuoli.

Descœudres, J-P., Sear, F. 1987. The Australian expedition to Pompeii *Rivisti di Studi Pompeiani* 1: 11-36.

Ehrhardt, W. 1988. *Casa del Orso.* Band 2, Häuser in Pompeji. Deutsches Archäologisches Institut. Hirmar Verlag. München

Ehrhardt, W. 1998. *Casa di Paquius Proculus.* Band 9, Häuser in Pompeji. Deutsches Archäologisches Institut. Hirmar Verlag. München

Eschebach, H. 1979. *Die Stabianer thermen in Pompeji.* Deutches Archäologisches Institut. Band 13 Berlin. De Gruyter & Co.

Eschebach, L. 1993. *Gebaudeverzeichnis und Stadtplan der antiken Stadt POMPEI.* Wien: Bohlau Verlag

Franklin, J. L.Jr. 1991. Literacy and the parietal inscriptions of Pompeii In:Humphrey, J. H.(Ed.) *Literacy in the Roman World.* Ann Arbor

Fröhlich, T. 1996. *Casa della Fontana Piccola.* Band 8, Häuser in Pompeji. Deutches Archäologisches Institut. Hirmar Verlag. München

Gierow, M. S. 1994. *Casa del Granduca (VII.4.56) und Casa dei Capitelli Figurati (VII.4.57).* Band 7, Häuser in Pompeji. Deutsches Archäologisches Institut. Hirmar Verlag. München

Gierow, M. S. 2000. *Casa delle Parete Nera (VII.4.58-60), Casa delle Forme di Creta (VII.4.61-63).* Band 10, Häuser in Pompeji. Deutsches Archäologisches Institut. Hirmar Verlag. München

Gusman, P.c1900. *Une ville antique sous les cendres: Pompéi* Paris: Société Française D'Éditions d'Art.

Hobson, B. 2004. *Domestic Toilets in Pompeii: Human waste disposal practices over four centuries in Regio VI, Insula 1.* Unpublished M. Phil. Thesis. University of Bradford.

Horne, John Fletcher 1895. *The buried cities of Vesuvius, Herculaneum and Pompeii* London: Hazell, Watson and Viney.

Jansen, G. C. M. 1992. Water systems and sanitation in the Houses of Herculaneum *Mededelingen van het* Nederlands *Institut te Rome 50*, 145-166.

Jansen, G. C. M. 1993. Paintings in Roman toilets In: Moormann, E. M. (ed.) Functional and spatial analysis of wall painting: 29-33 Leiden: Stichting Babesch.

Jansen, G. C. M. 1994. Sewers and tap water as urban innovations at Herculaneum In:XIVè Congrés Internacional D'Arqueologia Clàssica:218-20: Tarragona.

Jansen, G. 1997. Private toilets in Pompeii: Appearance and operation In: Bon, S. E., Jones, R. (eds.) Sequence and Space in Pompeii Oxbow Monograph 77:121-134 Oxford: Oxbow Books.

Jansen, G. C. M. 2000a. Systems for the disposal of waste and excreta in Roman cities. The situation in Pompeii, Herculaneum and Ostia In: Raventós, X. D., Remolà, J-A. (eds.) Sordes urbis. La eliminación de residuos en le ciudad Romana: 37-49 Roma: L'Erma di Bretschneider.

Jansen, G. C. M. 2000b. Studying Roman hygiene: the battle between the 'optimists' and the 'pessimists'. In: Jansen, G. C. M. (ed.) Cura Aquarum in Sicilia: 275-9 Leiden: Stichting Babesch.

Jansen, G. C. M. 2001. Water pipe systems in the houses of Pompeii. Distribution and use. In: Koloski-Ostrow, A. O. (Ed.) Water use and hydraulics in the Roman city AIA Colloquia and Conference Papers 3: 27-40 Boston.

Jansen, G. C. M. 2003 Social distinctions and issues of privacy in the toilets of Hadrian's Villa. *Journal of Roman Archaeology* 16: 137-152.

Jashemski, W. F. 1979. *The Gardens of Pompeii, Herculaneum and the Villas Destroyed by Vesuvius.* New York: Caratzes Bros.

Ling, R. 1997. *The Insula of the Menander at Pompeii.* Volume I: The Structures. Oxford: Clarenden Press

Love, M. 2007. *Analysis of Calcareous deposits from the down pipes of first century Pompeii* Unpublished Undergraduate Dissertation. University of Bradford.

Mau, A. (Trans. Kelsey, W.) 1902. *Pompeii: Its Life and Art.* New York: Macmillan.

Michel, D. 1990. *Casa dei Cei.* Band 3, Häuser in Pompeji. Deutches Archäologisches

Institut. Hirmar Verlag. München

Mygind, H. 1921 Hygienische Verhältisse im alten Pompeji Janus 25:251-383

Neudecker, R 1994 *Der Pracht die Latrine* Műnchen: Verlag Dr. Friedrich Pfeil

Packer, J. 1978. Inns at Pompei *Cronache Pompeianae 4*:5-53.

Pappalardo, U. 2001. *La Descrizione di Pompei per Guiseppe Fiorelli (1875)* Napoli: Massa Editore.

Parslow, C. 2000. The hydraulic system in the *balneum venerium et nongentum* of the Praedia Iuliae Felicis in Pompeii. In:Jansen, G. C. M. (Ed.) *Cura Aquarum in Sicilia:* 201-9 Leiden: Stichting Babesch.

Pirson, F 1999. *Mietwohnungen in Pompeji und Herculaneum* Verlag Műnchen: Dr Friedrich Pfeil.

Scobie, A. 1986. Slums, Sanitation, and Mortality in the Roman World. *Klio 68:399-433.*

Sear, F. 2004. Cisterns, Drainage and Lavatories in Pompeian Houses, Casa del Granduca PBSR Vol.LXXII:125-166

Seiler, F. 1992. *Casa degli Amorini Dorati.* Band 5, Häuser in Pompeji. Deutches Archäologisches Institut. Hirmar Verlag. München

Sogliano, A. 1900. Pompei – relazione degli scavi eseguiti durante il mese di novembre 1900 *Notizie degli scavi 1900:* 584-603.

Stemmer, K. 1992. *Casa dell' Ara Massima.* Band 6, Häuser in Pompeji. Deutsches Archäologisches Institut. Hirmar Verlag. München

Strocka, M. 1984. *Casa dei Principe di Napoli.* Band 1, Häuser in Pompeji. Deutsches Archäologisches Institut. Hirmar Verlag. München

Strocka, M. 1991. Casa del Labirinto. Band 4, Häuser in Pompeji. Deutsches Archäologisches Institut. Hirmar Verlag. München

Thédenat, H. 1910. *Pompei* Librairie Paris : Renouard.

Wilson, A 2000. Drainage and sanitation. In: Wikander, O. (Ed.) *Handbook of Ancient Water Technology* Vol 2:141-179. Leiden: Brill

CLASSICAL REFERENCES

Celsus: *De Medicina* 1971. (Trans. W. G. Spencer). Loeb Classical Library 292, 304, 336. Cambridge, Mass.: Harvard University Press.

C.I.L. Corpus Inscriptionum Latinarum

Frontinus: *The Stratagems: The Aqueducts of Rome.* (Trans. Charles E. Bennett). Loeb Classical Library 174 Cambridge, Mass.: Harvard University Press

Pliny *Natural History* 1963. (Trans.Jones, W. H. S.). Loeb Classical Library. Cambridge.

Mass.: Harvard University Press.

Strabo: *Geography* 1929 (Trans. Page, T. E., Capps, E., Rouse, W.H.D). Loeb Classical Library. Cambridge. Mass.: Harvard University Press.

Strabo *The Geography of Strabo II* 1969 (trans. Jones, H. L.) Loeb Classical Library. Cambridge. Mass.: Harvard University Press.

Vitruvius :*The Ten Books on Architecture.* (Trans. Morris Hicky Morgan) New York: Dover Publications

COLOUR PICTURES

VII. 2. 18

VII. 6. 30. 37 After cleaning: 2007

VII. 1. 44, 45 room d

VIII.3.16 Niche latrine

IX.2.15 Upper storey Latrine in outer wall (balcony) with down pipe

VII.1.36 Latrine in bakery

VI.5.8,20

I.11.10

VI.6.1

VI.11.6,13 Latrine with foot rest. Drain through wall

V.3.10 Down pipe draining into small pipe drain of latrine

Latrine with two pedestals

I.1.10 Down pipe

V.1.11,12 Double seat latrine with down pipe in masonry

VI.14.30,31,32 Down pipe

VII. 2. 42 Downpipe

VII. 3. 16 Down pipe

VII. 3. 35 Down pipe

VII.9.50 Internal down pipe

VII. 12. 13 Down pipe

IX.2.23 Down pipe in buttress, room

IX.3.25 Latrine with down pipe

VII.4.38

IX. 3. 14 Down pipe encased in masonry

www.ingramcontent.com/pod-product-compliance
Lightning Source LLC
LaVergne TN
LVHW072121210926
848482LV00028B/709